The Urban Book Series

Editorial Board

Margarita Angelidou, Aristotle University of Thessaloniki, Thessaloniki, Greece

Fatemeh Farnaz Arefian, The Bartlett Development Planning Unit, UCL, Silk Cities, London, UK

Michael Batty, Centre for Advanced Spatial Analysis, UCL, London, UK

Simin Davoudi, Planning & Landscape Department GURU, Newcastle University, Newcastle, UK

Geoffrey DeVerteuil, School of Planning and Geography, Cardiff University, Cardiff, UK

Jesús M. González Pérez, Department of Geography, University of the Balearic Islands, Palma (Mallorca), Spain

Daniel B. Hess⒟, Department of Urban and Regional Planning, University at Buffalo, State University, Buffalo, NY, USA

Paul Jones, School of Architecture, Design and Planning, University of Sydney, Sydney, NSW, Australia

Andrew Karvonen, Division of Urban and Regional Studies, KTH Royal Institute of Technology, Stockholm, Stockholms Län, Sweden

Andrew Kirby, New College, Arizona State University, Phoenix, AZ, USA

Karl Kropf, Department of Planning, Headington Campus, Oxford Brookes University, Oxford, UK

Karen Lucas, Institute for Transport Studies, University of Leeds, Leeds, UK

Marco Maretto, DICATeA, Department of Civil and Environmental Engineering, University of Parma, Parma, Italy

Ali Modarres, Tacoma Urban Studies, University of Washington Tacoma, Tacoma, WA, USA

Fabian Neuhaus, Faculty of Environmental Design, University of Calgary, Calgary, AB, Canada

Steffen Nijhuis, Architecture and the Built Environment, Delft University of Technology, Delft, The Netherlands

Vitor Manuel Aráujo de Oliveira⒟, Porto University, Porto, Portugal

Christopher Silver, College of Design, University of Florida, Gainesville, FL, USA

Giuseppe Strappa, Facoltà di Architettura, Sapienza University of Rome, Rome, Roma, Italy

Igor Vojnovic, Department of Geography, Michigan State University, East Lansing, MI, USA

Claudia Yamu, Department of Spatial Planning and Environment, University of Groningen, Groningen, Groningen, The Netherlands

Qunshan Zhao, School of Social and Political Sciences, University of Glasgow, Glasgow, UK

The Urban Book Series is a resource for urban studies and geography research worldwide. It provides a unique and innovative resource for the latest developments in the field, nurturing a comprehensive and encompassing publication venue for urban studies, urban geography, planning and regional development.

The series publishes peer-reviewed volumes related to urbanization, sustainability, urban environments, sustainable urbanism, governance, globalization, urban and sustainable development, spatial and area studies, urban management, transport systems, urban infrastructure, urban dynamics, green cities and urban landscapes. It also invites research which documents urbanization processes and urban dynamics on a national, regional and local level, welcoming case studies, as well as comparative and applied research.

The series will appeal to urbanists, geographers, planners, engineers, architects, policy makers, and to all of those interested in a wide-ranging overview of contemporary urban studies and innovations in the field. It accepts monographs, edited volumes and textbooks.

Indexed by Scopus.

More information about this series at https://link.springer.com/bookseries/14773

Susanna Clemente

Reconstructing Theatre Architecture

The Developing Process of the Modern Space for the Show

Susanna Clemente
Dipartimento di Architettura e Progetto
Sapienza University of Rome
Rome, Italy

ISSN 2365-757X   ISSN 2365-7588   (electronic)
The Urban Book Series
ISBN 978-3-030-89967-7   ISBN 978-3-030-89968-4   (eBook)
https://doi.org/10.1007/978-3-030-89968-4

© Springer Nature Switzerland AG 2022
This work is subject to copyright. All rights are reserved by the Publisher, whether the whole or part of the material is concerned, specifically the rights of translation, reprinting, reuse of illustrations, recitation, broadcasting, reproduction on microfilms or in any other physical way, and transmission or information storage and retrieval, electronic adaptation, computer software, or by similar or dissimilar methodology now known or hereafter developed.
The use of general descriptive names, registered names, trademarks, service marks, etc. in this publication does not imply, even in the absence of a specific statement, that such names are exempt from the relevant protective laws and regulations and therefore free for general use.
The publisher, the authors and the editors are safe to assume that the advice and information in this book are believed to be true and accurate at the date of publication. Neither the publisher nor the authors or the editors give a warranty, expressed or implied, with respect to the material contained herein or for any errors or omissions that may have been made. The publisher remains neutral with regard to jurisdictional claims in published maps and institutional affiliations.

This Springer imprint is published by the registered company Springer Nature Switzerland AG
The registered company address is: Gewerbestrasse 11, 6330 Cham, Switzerland

There is no place for theatre to actually settle down: its 'journey' allows the most adventurous and casual meetings made completely surreal by the hieratic nature of its demeanor.[1]

Theatre is interpretation and representation. Theatre is reality, a certain and tangible realization of interpretation and representation. Theatre is the form of all the arts. Architecture interprets and represents space, just as Theatre is a synthetic realization of idea.

Theatre Architecture as a space for the show determines a passage of a higher order, it can be understood as a representation for and of a representation.

The identification of the space for and of the show comes strictly from the needs and is always linked to the specific characters of representation. Hence, the thought of the continuous journey by Manfredo Tafuri quoted in the opening.

---

[1] *Per il teatro non esiste alcun luogo dove effettivamente depositarsi: il suo "viaggio" permette gli incontri più avventurosi e casuali resi del tutto surreali dalla ieraticità del suo contegno.* Tafuri M (1982) L'éphémère est éternel. In: Brusatin M, Prandi A (ed) Aldo Rossi. Il teatro del mondo, CLUVA, Venezia, p. 148.

To Maurizio Varamo

Acknowledgments

I really would like to thank first Professor Giuseppe Strappa for following and encouraging me, starting with the PhD in Architecture and Construction of Sapienza, which he was directing at the time, during the years, throughout the research path. His teachings, his studies on the formative process of the city, on the concept of "knotting" are the basis of the development of this study.

This work is also the result of a great passion for theatre, which I have cultivated since I was a child, and in which I had the opportunity to try myself at the Scenography Workshops of the Teatro dell'Opera di Roma, directed by Maestro Maurizio Varamo, who tragically died in 2019.

I owe my direct knowledge of Italian theatres to the attention and availability of the staff of the Teatro dell'Opera di Roma, the Teatro San Carlo, the Teatro alla Scala, the Teatro La Fenice, and again the Teatro Comunale di Ferrara, the Teatro dei Filodrammatici, the Teatro Piccolo di Milano, the Teatro San Babila and the Spoleto Festival dei Due Mondi.

Archival research was made possible thanks to the support received from the Academy of Denmark, the American Academy, the Biblioteca Comunale Ariostea, the Bibliotheca Hertziana, the Biblioteca teatrale SIAE del Burcardo, the Biblioteca Teresiana, the British School at Rome, the Centro Archivi di Architettura, MAXXI, the Germanic Archaeological Institute of Rome, the state and municipal historical archives of Ferrara, Mantua, Milan, Rome, Venice.

Finally, the contact and discussion with the teachers and organizers of conferences, symposia and events at the UAL - University of the Arts London, Prague Quadrennial and World Stage Design was essential to acquire knowledge and research ideas.

Presentation
The Theatre as an Urban Knotting
by Giuseppe Strappa

This work by Susanna Clemente, "Reconstructing Theatre Architecture", contains a flood of ideas often hidden among many examples and suggestive descriptions. It is therefore quite difficult to present it in a few lines.

I would say, briefly, that one of its main centers of interest consists in the fertile idea of considering modern theatre not as the result of the transformation of ancient types, according to a narrative constantly fed by history manuals, but that it was born and developed from urban life itself, inside the continuous flow of experiences that runs through the European city and that is "knotted" in a destined place, revolving around a dense space in transformation.

As you can expect, the study involves a series of terms usually linked to the processes that form the city, the "urban fabric", above all, as a society of buildings that establish a common law, and the "urban knot" as the intersection of architectural continuities.

The real premise of the formation of modern theatre is, in fact, the use of open spaces, the evolution of theatrical representation as a specialization of popular forms of entertainment (religious recitations, secular festivals, jousting) that take place in open public places, surrounded by the urban fabric, real or ephemeral.

At the base of the first forms of entertainment, there was always the process of binding, the idea of tightening an open space, transforming it into a place where the vital flows of the neighborhood or the city meet. A place that could later be permanently covered, becoming built architecture.

This idea of an architectural gesture that magically transforms space through actions rather than through stable structures, is extendable: the representation could also take place in the free spaces of existing buildings, rotate and wrap around its courtyards, for example, on which it lay a simple velarium that transforms an empty space into a transitory architectural knot, as in the palace of Cardinal Raffaele Riario, where already in the fifteenth century spectacles were performed in the courtyard, temporarily transformed into a large hall through canvas covering.

The original form of the Elizabethan theatre itself, in this sense, can be considered as the translation into a building of this urban custom, where the show was initially represented in open spaces surrounded by a fabric of wooden boxes. In the well-known drawing by Johan deWitt of the Swan theatre in London (1596), the public space of the parterre/square appears crowded of spectators, standing or on makeshift seats, bounded by the "private" structures of the boxes, built with the logic of a building aggregate.

In fact, the play was a part of the more general spectacle of the crowd that met in the shared space. A perimeter of structures inhabited for a few hours by the participants in the event therefore surrounded the central space which, initially opened, once covered ended up forming a spatial center, transforming an urban fabric into a building. The covering of the stall, building a large central hall, will give rise to a type of theatre that will remain in use for centuries.

For these reasons, it seems to me that one of the fundamental notions underlying the author's work is that of "knotting". A new and very fertile notion in every field of architecture, which also allows us to read the transformation of the structures for the show as a progressive search for a unitary space within a part of the city changed into architecture by specialization.

In general, "knotting" indicates the outcome of the constructive gesture of connecting different elements of a structure together in order to form a spatial node within an architectural or urban organism, tying it to the surrounding structures, usually consisting of a series of rooms. Many types of past and modern buildings are formed by knotting, generated by the dialectic between fence and roof, between serial and organic structures.

Complementary to the idea of knotting is, in fact, that of the act of fencing. The fence is the result of the act of wrapping a limited portion of space with a continuous structure, triggering its own functional/symbolic machine, linked to the stratification of the forms through which man has experienced a concluded space, generating a code, within certain limits, constant and universal.

The formative processes of the enclosure are therefore linked to the concepts of centrality and peripherality, of nodality and antinodality which constitute a fundamental key to understanding the character of architectural organisms.

In this interpretation, the routes and axes play a fundamental role, the substantial difference between them being contained in the different intentionality that generates them: the first produced by the spontaneous use of the anthropized space, the second fruit of the architectural reflection on the principles that regulate the life of the built organism. Having a direction, the axis (complementary to the dividing lines that separate the spaces) establishes a succession of elements, orienting the reading from the initial to the terminal structures. For this reason, the elements encountered take on a different value and meaning. This finding implies that the hierarchization of the elements inside the enclosure does not simply respond to a geometric logic, but is linked to the ways in which man uses and knows space, and to the time in which the process of its knowledge and use takes place. The author uses these notions in a new way, applying them to figurative contexts that are usually approached with only the tools of perception: to the representation of the Passion of Villingen, to that of Valenciennes, to one of the first theatre drawings, *The Castle of Perseverance*, the martyrdom of Saint Apollonia painted by Fouquet and many others.

Presentation The Theatre as an Urban Knotting by Giuseppe Strappa

It seems important to me to note how, in a cultural context dominated by the present, the exemplary work of Susanna Clemente is particularly valuable to rediscover the formative processes that explain how what we see, the spaces we inhabit, are the product of a succession of knowledges.

This method of seeing forms as matter in transformation has a great significance because it allows us to read the built reality not only for what it is but for what it could be. It is therefore useful to the architectural project as well as to that of the stage space.

I believe that this consideration is fundamental in the work of the author, directly involved in the theatrical activity that fascinates her not only as a scholar but also for her direct interest in the search for a new relationship between city and representation. Relationship that she constantly intends to "prove".

Susanna Clemente began this work in the Ph.D. in Architecture and Construction of Sapienza, which I was directing at the time. The centre of our work, and also its originality, consisted precisely in seeking the demonstration: how what we see today is the effect of forming phases read in the built landscape, and whose innovative interpretation can give pioneering results in the contemporary project.

Demonstration is a central theme today. In a society of consumers in which architects do not feel the duty to prove things, it also has an ethical content, it represents a value.

I believe that in her important study, Susanna Clemente also used this contribution to her formation as a researcher in a rigorous way, arranging the subject matter according to a scientific, but also original and creative method. And this is a source of great satisfaction for me.

Introduction

Representation and graphic representation

Theatre is very similar to architecture because it concerns a story; its beginning, its development and its conclusion. Without a story there is no theatre and there is no architecture. It is also moving that each one experiences a small part of his own.[1]

Before getting to the heart of the research, we want to draw attention to those aspects that were considered fundamental for conducting the study, valid in all respects as an assumption.

Undertaking the above-mentioned path, the first problem is related to the absence of theatres for most of the medieval and modern era. *Reconstructing Theatre Architecture*, then, to underline the developing process of the space for the show, to which is added the lack of direct evidence of the modern space of representation. Generally, few modern theatres have come down to us in the whole national territory, and almost all dating back to no earlier than the fifteenth-sixteenth centuries.

In Konigson's *The Space of the Theatre in the Middle Ages*, we read, almost in the opening, that theatre *is simultaneously inside the city and outside it, connected to the flow of human activity and opposite of it.*[2] In Medieval Times, there is no specific dedicated space indeed, primarily theatrical. Theatre finds its dimension by combining with the existing, with the built environment, starting from the liturgical drama inside the church up to the theatre of the mysteries, the farces, the entrances, which instead are based in the city as a whole. The representations in the squares are so rooted in the tradition of the places that even when the choral of the medieval municipality is abandoned to move on to the individuality of the spectacle of the lords and courts, the open or closed courtyards, the halls the parties are covered like

---

[1] *Il teatro è molto simile all'architettura perché riguarda una vicenda; il suo inizio, il suo svolgimento e la sua conclusione. Senza vicenda non vi è teatro e non vi è architettura. È anche commuovente che ognuno viva una sua piccola parte.* Rossi A (1990) Autobiografia scientifica. Pratiche, Parma: 58.

[2] *contemporaneamente dentro la città e all'esterno di essa, connesso al flusso dell'attività umana e all'opposto di esso.* Konigson E (1990) Lo spazio del teatro nel medioevo. La Casa Usher, Lucca: 37.

Fig. 1 Hans Memling—The passion of Turin [Wikipedia]

real covered squares. Since there is no certain, univocal identification of the space, this is never presented as unified, but rather discontinuous.

Two works by the painter Hans Memling are often cited as emblems, plastic representations of medieval scenic principles, *The Passion of Turin* (1471) and *The Seven Joys of Mary*. In particular, the first (Fig. 1), quoted several times starting by Cavalcaselle and following by Magagnato,[3] is recognized as a product of the dramatic influence.

The scenes of the last acts of Christ's life unfold in the landscape, in a modern city. The drama is then placed, as we have said, in the context of the urban fabric. However, the observer's gaze can encompass all the episodes, the different times and the different actions in a single gesture. The narrative begins at the left vertex, with the entry into Jerusalem, followed by the expulsion of the merchants from the temple against the background of a porticoed square; the betrayal of Judas and the Last Supper are confined to internal environments through windows and arches, architectural expedients to link together, contextualizing them, the different events. In the foreground, there are events outside and inside the city walls: the garden of Gethsemane, the capture, the presentation of Christ to Pilate, the construction of the cross, the denial of Peter, the scourging, up to the Via Crucis. From this, we go up towards the right vertex passing through the deposition in the sepulcher and the Resurrection, the Noli me tangere and further away the meeting on the road to Emmaus. It should be noted that not only in the architectural contexts but also in the external settings each group of characters, protagonists of the events, is similar to a tableau vivant. Even the diversified lighting of the scenes takes on a symbolic value

---

[3] Magagnato L (1954) Teatri italiani del Cinquecento. Neri Pozza Editore, Venezia: 14–16.

Introduction

Fig. 2 Hans Memling—The seven joys of Mary [Wikipedia]

that facilitates the reading of the painting. On the left the darkness, the dim light of the torches, on the right the resurrection, the incident light to emphasize the square and the hill, dramatic highlights.

We are faced with a rhythmic image, of process, of development, a complex way of staging and enjoying the show, which necessarily requires an open, adaptable, multifunctional space. Certainly, the square is a privileged place, it becomes an image not only of the city but also the cosmos. The principle of simultaneity—in *The Seven Joys of Mary* all the moments of the story and all the scenographic elements are painted on the canvas from Nativity to Pentecost—is therefore completely dissimilar and indeed does not accord with the "processional" character typical of some events in subsequent periods (Fig. 2). Even in the famous Bayeaux tapestry, this character is only apparent, since the story consists of continuous references and very far from a linear development.

The different, multiple medieval scenes are not placed in space according to a narrative scheme. Indeed, the narrative itself involves a going on and back to the same places, or even more settings can be created in a single area (like the internal rooms of buildings without specifications or recognizable attributes in *The Passion of Turin*). In the analyzed works, the movement itself from one scene to the next is not linear, unidirectional. The scenes are instead placed in space following a symbolic tradition, a hierarchical order, but also the needs of the staging.

In the first chapter of the research, orientation will constitute the first step in the process of creating the space for the show, the primary form of the construction act.

As seen in the beginning, the absence of spaces for the show dating from the Middle Ages onwards and still existing today increases the meaning and importance of their representations. Representations that, however, cannot help but bring the theatrical representation, the dynamism and the complexity of the action within them. "Question" which reaches "intact" up to the eighteenth century with the théatre de Besançon by Ledoux and beyond. The engraving was supposed to show the stage set up for the inauguration of the theatre. The public theatre enters, is extended into the city through the stage and is at the same time closed by the stage, delimited in its artifice.

City—but also square, indicating the part for the whole—or even architecture as theatre, and theatre as city, are very recurrent combinations; there is often a vague resemblance of this two concepts containing ideas, actions, spaces and architectures. The link, on the other hand, is much deeper, as both are subject to the same evolutionary process. Here, therefore, is the analysis of Rossi's opening observation.

Representation, whether graphic or dramaturgical, is an active mediation tool that allows us to externalize our ideas, to give them independence and autonomy in a context that is structurally similar to real space, through a series of conventions or norms. In Latin, there are three terms that translate the Italian 'representation': *pictura*, representation of a drawing, *imago*, representation of an idea and *spectaculum*, theatrical performance.

Quoting William of Ockham: *Representing has several senses. In the first place, this term means what with one knows something and in this sense knowledge is representative, [...]. Secondly, to represent means knowing something, known which another thing is known; and in this sense the image represents what the image is of. Thirdly, it is meant to represent causing knowledge to the way in which the object causes knowledge.*[4] An interesting parallel can be identified by relating the three definitions with the Vitruvian *ichnographia*, *orthographia* and *scenographia*, where ichnographia means the imprint left by the building on the ground, the plan, the most general idea generating architecture; by orthographia, we mean the drawing of the elevations, of what comes from the idea, and therefore is an image of this; finally, scenography is the representation of the object closest to reality, to the essence of it.

At the end of this brief excursus aimed at underlining the various levels of representation, attention is therefore drawn to the particularity of representing an object, such as the theatrical place, which is itself intended for representation.

The theme of "intermediality", a neologism brought to the fore by contemporary American research, absent from the first modern lexicon, is introduced below, in support of what has been said about the problematization of graphic representation. It is interesting to observe how theatre is by definition intermedial and to understand how today communication through the media has become a self-aware condition of theatre as well as architecture, a means of expression and exchange. It was precisely in the fifteenth century that the diversification of the media and the growing dialectic between the modalities of art began to be recorded. The role of drawing has been recognized as central in this theorization process. Towards the middle of the sixteenth century, Italian theorists confirmed design as the godmother, the generator and the synthesis of all disciplines. The preparatory drawings also began to have great influence at the time, although they were not, however, the only means available. Think of the models, the prints (Fig. 3), etc. The idea is therefore first verified and then

---

[4] *Rappresentare ha parecchi sensi. In primo luogo si intende con questo termine ciò con cui si conosce qualcosa e in questo senso la conoscenza è rappresentativa, [...]. In secondo luogo si intende per rappresentare il conoscere qualcosa, conosciuta la quale si conosce un'altra cosa; e in questo senso l'immagine rappresenta ciò di cui è l'immagine, nell'atto del ricorso. In terzo modo si intende per rappresentare il causare la conoscenza al modo in cui l'oggetto causa la conoscenza.* Apollonio FI, (2009) Della rappresentazione e del suo doppio. In: Braghieri G (ed), Architettura 32. Il luogo della rappresentazione. CLUEB, Bologna: 31–32.

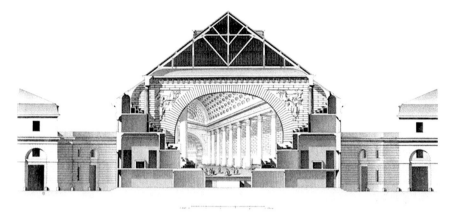

Fig. 3 Claude Nicolas Ledoux—Théatre de Besançon [Wikipedia]

transmitted through the main means of drawing, which, however, despite being the most concise and capable of understanding the various aspects within it, the multimedia of theatre, is not able to be exhaustive. You will have to wait until the present day to successfully go beyond the purely graphic representation, to achieve real-time sharing of the moving image.

Finally, further areas that guide the research are all sources improperly called indirect, therefore not built as the very few theatres of past centuries still standing, and not figurative like all the kinds of performances mentioned above. There are numerous descriptions in chronicles, accounting registers, property documents, etc., from which we will try to derive, in some cases, possible figurative interpretations.

In summary, before getting to the heart of the research, we wanted to draw attention to those aspects that were considered fundamental for conducting the study, valid in all respects as an assumption:

- the absence of built theatres for most of the medieval and modern eras;
- the absence of modern theatres still existing today (almost all dating back to no earlier than the fifteenth-sixteenth centuries);
- to rely on the representations of spaces for representation with all the limits or, on the contrary, all the strengths that a double jump produces;
- the importance of non-direct and non-representative sources and the attempt to draw from these unpublished images.

The main question of the research is the identification of the developing process of the modern space for the show. So far, the causes at the origin of the question itself have been briefly described as a premise: first of all, the void, the absence of evident links between classical theatre and Renaissance theatre, with an essential function of innovation represented by the entire medieval era, undoubtedly the most complex to reconstruct due to the scarcity of sources, direct and indirect; secondly, the intrinsic difficulty of representing and transmitting the space for representation.

The main objective of the research is to overcome the gap between the vestiges of classicism and modern theatre, whose main characteristics have typically established in our peninsula, and then spread first in the European and then in the Western context in general. Then we want to investigate those recurring elements, because they were successful, which led to the development of the theatrical type, studying its mutations through the historical construction, motivating the clear break found with the models of antiquity and at the same time highlighting the reasons for the diffusion and pre-eminence of the Italian model.

The research extends over a vast period of time, from the early Middle Ages to the present, highlighting the major aspects of mutual influence with the European panorama and emphasizing the uniqueness of the Italian results. Italy is the country where the largest number of historical theatres survive (Fig. 4), still in use and with traditional scenographic and show techniques: a heritage that must be preserved and, at the same time, in continuous evolution. The study is consequently largely focused on the so-called "Italian" theatre, however, meaning that this assumption is not limiting, but rather majority and coexisting with less represented and representative models on the national scene, which will nevertheless be described and related to each other and with the various stages of the evolutionary process.

The identification of a training process for the modern space for the show wants to open a new way of reading theatre as an architectural object, providing at the same time specific design guidelines to operate coherently, reaffirming the centrality and autonomy of theatrical architecture as a discipline operating in a multidisciplinary context par excellence.

Methodologically, the research was conducted according to the following phases:

- identification of the research topic and consequent study and revision of the essential bibliography, mostly consisting of treatises, modern and contemporary, national and European, with specific attention to the Italian model;
- tracking of the state of the art with the highlighting of the main orientation questions for the research;
- collection of data through direct and indirect sources, visits to major Italian theatres, literature and original documents in part unpublished through the consultation of the main libraries in the fields of architecture, history, history of art and entertainment as well as at municipal historical archives and state archives;
- transcription of data: recognition of the uniqueness of the Italian affair and circumscription of the case studies to the highest level and at least provincial realities;
- analysis of the case studies: identification of the key points of the process of forming the modern space for the show with consequent classification of each project within one or more phases of the evolutionary process;
- in-depth analysis of the evolution of the main typological characteristics identified downstream of the training process and of the prospects for intervention on and in the building;
- general conclusions and perspectives of the research.

Introduction

Fig. 4 The approximately 700 historical theatres in Italy [elaboration by the author]

European Route of Historic Theatres:

Dozens of historical Italian theatres analyzed are included in the itineraries of the ERHT Project, a European project that has involved and still involves 12 European countries and 16 different partners.

UNESCO sites where some of the historical Italian theatres analyzed are located:

Historic center of Rome, Mantua and Sabbioneta, City of Verona, Historic center of Urbino, Ferrara, city of Renaissance, and its Delta of Po river, the Royal Palace of the eighteenth century of Caserta, Historic center of Naples, Venice and its lagoon, Historic center of Siena, Historic center of Florence, the City of Vicenza.

Reading Guide

The text is divided into six chapters, each one signifying a phase of the developing process that we are trying to reconstruct. The theme is based on specific case studies, in the absence of a specific chronological order. In fact, the particular interpretative line adopted is what we want to highlight; it describes the composition of the theatrical space as an exquisitely urban process, as a product of our cities, and the turning points, which can be identified in different cases located in chronological moments, sometimes very distant.

Relevant episodes are detailed in the tables at the end of each chapter, with a summary of the results.

The first chapter defines the intrinsically changing character of theatrical architecture, paying attention to the primary constructive actions, belonging to each architectural organism, of fencing and covering.

The second chapter intends to analyze the emergence of the characteristics of autonomy and recognition of the space intended for the show through the study of temporary events.

Then, in the third one, the aspects of Vitruvianism, rediscovery, interpretation and invention, persistent not only in Palladian but also in Scamozzi's work, are investigated, clarifying the rediscovery and the relationship with the classical world.

In *Theatre as solidification*, *Theatre as recast* and *Theatre as urban knot,* the main phases of the process of specialization of the type were explored, up to the final chapter dedicated to the evolution of the characters of theatre.

The work is completed by documentary and iconographic apparatuses, as well as by the list of about seven hundred Italian historical theatres, also with the main data relating to location, time of construction and restoration and construction features. This analysis made it possible to focus on the uniqueness of the theme and the objectives of the research.

About This Book

Reconstructing Theatre Architecture is a study aimed at reconstructing the historical process at the base of any significant theatre architecture. The modern space for the show is no longer intended as a direct derivation from classical types, but as a product of the transformation of the urban fabric in our cities. The study, which extends over a vast period of time, from the Middle Ages to our days, is intended to overcome the gap between the heritage of classicism and modern theatre.

Analytically reconstructing the events starting from the mobile origin, temporary and often linked to the occasionality of festivals, and bearing in mind the contemporary treatises, the research wants to arrive in phases at the reasoned definition of the characters of theatre, in and from the palace, of theatre as an autonomous node, on architectural and urban scale, up to the most recent developments in the subject.

The knowledge of the developing process of the modern space for the show can provide specific planning guidelines to operate coherently in the built environment.

The study methodologically refers to an interpretation of architectural design as closely related to the intentional and critical identification of typological processes. Once the typological process has been derived from the built environment, the action of the designer must be adapted to the process itself. Reading the architectural project is, equivalently to designing, an essential critical moment, based on the pre-existences, whose structure must be recognized from time to time in its manifestations. Therefore, thinking of the perfect equality of reading and writing, the experiential, experimental and comprehension aspect plays an essential role for the proposition of an original interpretative line of the theatrical organism, which must be identified, and with it open new perspectives.

The text is enriched with in-depth data sheets, documentary and iconographic apparatuses, which help to focus on the uniqueness of the theme and the new reading method that is proposed.

Contents

Part I   The Developing Process of the Space for the Show

1   Urban Knots as Space and Representation 3
   1.1   The Palladian Education by Aldo Rossi 3
   1.2   Fencing and Covering 7
   1.3   Synchronic Variants: Central Stage and Circular Theatre 12
   1.4   Summary 14
   References 25

2   The Theatrical Organism 27
   2.1   Temporarines and Permanency of the Features of the Theatre
       Space 27
   2.2   The Representation of the Renaissance and Baroque Festival
       References 35
   2.3   Summary 40
   References 48

3   Theatre as Solidification 51
   3.1   Reinventing Classical Theatre 51
   3.2   Vitruvius as a Source of Modern Constructive Pragmatism 54
   3.3   The Court Theatre of Mantua and the Farnese Theatre
       of Parma Towards Modern Theatre 58
   3.4   Summary 66
   References 74

4   Theatre as Recast 75
   4.1   Old and New Theatres in Mantua 75
   4.2   Wonderful Anticipations and Surprising Delays
       by an Author/Reader: Fabrizio Carini Motta 80
   4.3   A City Interior: The Disguised Theatre by Adolfo Natalini 89
   4.4   Summary 91
   References 94

5	**Theatre as Urban Knot**	97
	5.1 Public Theatre with Boxes, Italian Theatre	97
	5.2 The Synchronic Variant of Court and Academic Theatre	101
	5.3 Main Influences, Inbound and Outbound, Towards the European Context	107
	5.4 The Specialization of Building Types	109
	5.5 Summary	114
	References	126
6	**The Evolution of Theatre Features**	129
	6.1 A Theatre as an Urban Knot Grafted in the Historical Fabric	129
	6.2 From the Graft of the New to the Recovery of the Existing	136
	6.3 Scenography as "Architecture of the Theatre Space"	147
	6.4 Summary	149
	References	157

Conclusions	159
Apparatuses	163
Glossary	215
List of Historical Theatres in Italy	217
References	363

List of Figures

Fig. 1   Hans Memling—The passion of Turin [Wikipedia]   xvi

Fig. 2   Hans Memling—The seven joys of Mary [Wikipedia]   xvii

Fig. 3   Claude Nicolas Ledoux—Théâtre de Besançon [Wikipedia]   xix

Fig. 4   The approximately 700 historical theatres in Italy
[elaboration by the author]   xxi

Fig. 1.1   Aldo Rossi—Teatro del Mondo   5

Fig. 1.2   *Cysat—The Passion of Lucern, plan,* [http://theaterhistoryo
nline.blogspot.com/2014/08/lucerne-passion-play.html]   11

CODE 1.01 – IL TEATRO DEL MONDO   15

CODE 1.02 – THE PASSION OF VILLINGEN   16

CODE 1.03 – THE PASSION OF VALENCIENNES   17

CODE 1.04 – THE CASTLE OF PERSEVERANCE   18

CODE 1.05 – THE MARTYRS OF ST. APOLLONIA   19

CODE 1.26 – COMPARATIVE TABLE   20

Fig. 2.1   Giovan Battista Aleotti—Ferrara, 1605, Stampa, 527 ×
740 mm, Dedicata a Francesco Borghese [Biblioteca
Comunale Ariostea, Serie XVI-63]   31

Fig. 2.2   Piazza municipale, Ferrara [photo by the author]   32

Fig. 2.3   Town hall, Ferrara—details [photo by the author]   33

Fig. 2.4   Piazza del Duomo in Spoleto during the Festival dei Due
Mondi (2018) [photo by the author]   38

CODE 2.01 – TEATRO ALL'APERTO   41

CODE 2.02 – TEATRO ALL'APERTO   42

CODE 2.03 – CORTILE DI CASA CORNARO   43

CODE 2.04 – TEATRO DI PALAZZO VECCHIO   44

CODE 2.05 – TEATRO CAPITOLINO   45

CODE 2.06 – PIAZZA DEL DUOMO A SPOLETO   46

CODE 2.27 – COMPARATIVE TABLE   47

Fig. 3.1   Andrea Palladio—Olympic Theatre in Vicenza   52

Fig. 3.2   Vincenzo Scamozzi—Teatro all'Antica
in Sabbioneta—exterior [photo by the author]   55

Fig. 3.3	Vincenzo Scamozzi—Teatro all'Antica in Sabbioneta—interior [photo by the author]	57
Fig. 3.4	Vincenzo Scamozzi—the scene of Teatro all'Antica in Sabbioneta [photo by the author]	58
Fig. 3.5	Palazzo Giardino in Sabbioneta, Sala dei Circhi: scenographic fresco, 1588 ca. [photo by the author]	59
Fig. 3.6	Rilievo di Giuseppe Bianchi del Teatro Ducale, pianta [Archivio di Stato di Mantova, Archivio Gonzaga, H. VIII.—Teatri, giuochi, feste, maschere e lotterie, Teatri 1669—1787, busta n. 3170 bis]	60
Fig. 3.7	Mappa dell'Imperial Regio Teatro di Corte in Mantova [Archivio di Stato di Mantova, estratta da: I.R. Delegazione, a. 1863, busta n. 3588]	61
Fig. 3.8	Ex Teatro Regio, plan with demolitions and recontructions [Archivio Storico Comunale di Mantova, categoria V, classe 3, articolo 1, busta n. 30]	62
Fig. 3.9	Ex Teatro Regio, section and plans with demolitions and reconstructions [Archivio Storico Comunale di Mantova, categoria V, classe 3, articolo 1, busta n. 30]	63
Fig. 3.10	Palazzo Ducale in Mantua—Piazza Castello	64
Fig. 3.11	Palazzo Ducale in Mantua—Cortile della cavallerizza and Museo Archeologico Nazionale di Mantova [photo by the author]	64
CODE 3.01 – TEATRO OLIMPICO	67
CODE 3.02 – TEATRO DI SABBIONETA	68
CODE 3.12 – TEATRO DI SABBIONETA	69
CODE 3.03 – TEATRO FARNESE DI PARMA	70
CODE 3.24 – COMPARATIVE TABLE	71
CODE 3.25 – COMPARATIVE TABLE	72
Fig. 4.1	Detail of the Urbis Mauntuae Descriptio by Bertazzolo (1628) with Teatro Vecchio highlighted [elaboration by the author]	76
Fig. 4.2	Spiegazione della Pianta del Teatro Vecchio con la Distinta delle Case contigue di ragione della Regio Ducal Camera di Mantova, che sarebbero necessarie d'aggregarsi al detto Teatro per li giuochi alfine di abilitare il detto Teatro a ricavarne un congruo prodotto [Archivio di Stato di Mantova, Archivio Gonzaga, H. VIII. Teatri, giuochi, feste, maschere e lotterie, busta n. 3170]	77
Fig. 4.3	Teatro Vecchio [Archivio di Stato di Mantova, Archivio Gonzaga, H. VIII. Teatri, giuochi, feste, maschere e lotterie, busta n. 3170]	78
Fig. 4.4	Place where Teatro Fedeli stood [photo by the author]	79

List of Figures		xxix

Fig. 4.5   Lettera di G. Arnoldi contenente la richiesta prodotta alla Giunta Municipale di Mantova di variazione dei prospetti per la realizzazione del Teatro Arnoldi [Archivio Storico Comunale di Mantova, Titolo XIV—Sicurezza Pubblica, Articolo 24° Teatri, 1882–1884, Suddivisione 6—Teatro Arnoldi, Fascicolo n.1, busta n. 1093] 81

Fig. 4.6   Stato di fatto del futuro Teatro Arnoldi—prospetti lungo via Osterie e via Storta [Archivio Storico Comunale di Mantova, Titolo XIV—Sicurezza Pubblica, Articolo 24° Teatri, 1882–1884, Suddivisione 6—Teatro Arnoldi, Fascicolo n.1, busta n. 1093] 82

Fig. 4.7   Corografia del Teatro Arnoldi [Archivio Storico Comunale di Mantova, Titolo XIV—Sicurezza Pubblica, Articolo 24° Teatri, 1882–1884, Suddivisione 6—Teatro Arnoldi, Fascicolo n.1, busta n. 1093] 83

Fig. 4.8   Progetto del Teatro Arnoldi—prospetti lungo via Osterie e via Storta [Archivio Storico Comunale di Mantova, Titolo XIV—Sicurezza Pubblica, Articolo 24° Teatri, 1882–1884, Suddivisione 6—Teatro Arnoldi, Fascicolo n.1, busta n. 1093] 84

Fig. 4.9   Parere favorevole della Giunta Municipale di Mantova al progetto presentato da G. Arnoldi [Archivio Storico Comunale di Mantova, Titolo XIV—Sicurezza Pubblica, Articolo 24° Teatri, 1882–1884, Suddivisione 6—Teatro Arnoldi, Fascicolo n.1, busta n. 1093] 85

Fig. 4.10   Progetto del Teatro Arnoldi—prospetti lungo via Osterie e via Storta, contenente la variazione richiesta dalla Giunta Municipale di Mantova [Archivio Storico Comunale di Mantova, Titolo XIV—Sicurezza Pubblica, Articolo 24° Teatri, 1882–1884, Suddivisione 6—Teatro Arnoldi, Fascicolo n.1, busta n. 1093] 86

Fig. 4.11   The block today [elaboration by the author] 87

Fig. 4.12   Adolfo Natalini—Teatro della Compagnia, axonometry [http://www.nataliniarchitetti.com/progetti/spettacolo/8401.html] 88

Fig. 4.13   Adolfo Natalini—Teatro della Compagnia, view of the sage, sketch of the audience [http://www.nataliniarchitetti.com/progetti/spettacolo/8401.html] 90

CODE 4.01 – TEATRO DELLA COMPAGNIA 92

CODE 4.21 – COMPARATIVE TABLE 93

Fig. 5.1   Vecchio e nuovo teatro da costruirsi, con in evidenza le demolizioni [Archivio di Stato di Roma, Camerale III (secoli XV–XIX), busta n. 2127] 98

Fig. 5.2	Teatro Sociale in the urban context with demolitions highlighted. The most visible improvements, from the urbanistic point of view, would have been the liberation of the narrow district of San Francesco, the enlargement of the district of Concole, the rectification of the outlet of the district of Pradella on the square, and the disappearance of the alley of Sant'Antonio, which ran from the height of the Agnello district to the Folengo square then occupied by the Oratory of Sant'Antonino, as well as some houses	101
Fig. 5.3	Progetto per l'allargamento del Pronao—Teatro Sociale [Archivio Storico Comunale di Mantova, Titolo XIV—Sicurezza Pubblica, Articolo 24° Teatri, 1824—1831, Suddivisione 2—Teatro Sociale, busta n. 1085]	102
Fig. 5.4	Teatro Sociale—plan [elaboration by the author]	102
Fig. 5.5	Antonio Bibbiena—Teatro Scientifico dell'Accademia [photo by the author]	104
Fig. 5.6	Luigi Vanvitelli—Teatro di Corte [photo by the author]	106
Fig. 5.7	Palazzina Liberty, Milano	109
Fig. 5.8	Marco Zanuso—Il Piccolo Teatro Grassi and Il Piccolo Teatro Strehler in Milano [photo by the author]	110
Fig. 5.9	Teatro San Babila [photo by the author]	111
Fig. 5.10	Teatro Studio, Milano [photo by the author]	112
Fig. 5.11	Marco Zanuso—Il Piccolo di Milano [photo by the author]	113
CODE 5.01 – TEATRO TOR DI NONA		115
CODE 5.02 – TEATRO COMUNALE DI FERRARA		116
CODE 5.03 – TEATRO DELLA PERGOLA		117
CODE 5.04 – TEATRO DI CORTE		118
CODE 5.05 – TEATRO CARIGNANO		119
CODE 5.06 – PICCOLO TEATRO GRASSI		120
CODE 5.07 – PICCOLO TEATRO STUDIO		121
CODE 5.08 – PICCOLO TEATRO STREHLER		122
CODE 5.21 – TEATRO TOR DI NONA		123
CODE 5.29 – COMPARATIVE TABLE		124
Fig. 6.1	Carlo Barabino-Internal plant of the theatre [https://archit ettura.unige.it/e-books/berlendis/berlen_3.htm]	130
Fig. 6.2	Carlo Barabino—Hymnographies of the new Teatro Carlo Felice [https://architettura.unige.it/e-books/berlendis/ber len_3.htm]	131
Fig. 6.3	Carlo Scarpa—schizzo di studio [Centro Archivi di Architettura—MAXXI, Fondo Carlo Scarpa, Attività Professionale, Progetti e incarichi professionali, 214: Progetto Teatro Carlo Felice, Genova (1963–1981), n.18576]	132

List of Figures xxxi

Fig. 6.4	Carlo Scarpa—studio del foyer [Centro Archivi di Architettura—MAXXI, Fondo Carlo Scarpa, Attività Professionale, Progetti e incarichi professionali, 214: Progetto Teatro Carlo Felice, Genova (1963–1981), n.18673R]	133
Fig. 6.5	Carlo Scarpa—studio del foyer [Centro Archivi di Architettura—MAXXI, Fondo Carlo Scarpa, Attività Professionale, Progetti e incarichi professionali, 214: Progetto Teatro Carlo Felice, Genova (1963–1981), n.18708]	134
Fig. 6.6	Main views of Turin drawn in perspective and carved in copper by the architect Giambattista Borra—The eighteenth-century theatre [https://www.teatroregio.torino.it/scopri-il-regio/storia]	137
Fig. 6.7	The plan of the eighteenth-century theatre and that of the contemporary theatre by Carlo Mollino [elaboration by the author]	138
Fig. 6.8	Luigi Caccia Dominioni—Teatro dei Filodrammatici [photo by the author]	139
Fig. 6.9	Teatro Alla Scala [photo by the author]	140
Fig. 6.10	Graph showing the distribution in time, by century of construction, and in space, by region, of the approximately 700 historic Italian theatres; note in particular the great development that took place in Emilia-Romagna, Lombardy, Veneto, Tuscany and Marche, in line with what has been described so far; the data are obtained from the "List of historical theatres in Italy", created as part of the research and reported in the appendix [elaboration by the author]	141
Fig. 6.11	Teatro Alla Scala [photo by the author]	142
Fig. 6.12	Teatro San Carlo [photo by the author]	143
Fig. 6.13	Teatro La Fenice—exterior [photo by the author]	145
Fig. 6.14	Teatro delle 6, ex Palazzo della Signoria in Spoleto [photo by the author]	146
Fig. 6.15	Teatro Menotti [photo by the author]	146
Fig. 6.16	Teatro Caio Melisso [photo by the author]	147
Fig. 6.17	Graph showing the distribution by region of historic Italian theatres for which architectural interest is recognized, of theatres that are World Heritage Sites or listed in Unesco sites and finally of theatres part of the European Route of Historic Theatres; the data are obtained from the "List of historical theatres in Italy", created as part of the research and reported in the appendix [elaboration by the author]	148

Fig. 6.18 Graph showing the distribution by region of historical
Italian theatres subjected over time to interventions
of grafting the new, recovery of the existing, conservation,
restoration and philological reconstruction; the data are
obtained from the "List of historical theatres in Italy",
created as part of the research and reported in the appendix
[elaboration by the author] 149
CODE 6.01 – TEATRO CARLO FELICE 150
CODE 6.21 – TEATRO CARLO FELICE 151
CODE 0.00 – COMPARATIVE TABLE 152

Part I
The Developing Process of the Space for the Show

Chapter 1
Urban Knots as Space and Representation

Origin and Development of Entertainment Facilities

Abstract The chapter deals with the birth of the modern space for the show, it explores the most significant aspects starting from case studies dating back to the final phases of the medieval era and the dawn of Renaissance, as well as to recent projects, of contemporary times, which testify the act, constantly present and timeless, of generating the complex organism that theatre is. In particular, the in-depth analysis is divided into three main parts, of which the first moves from Aldo Rossi's Teatro del Mondo to analyze the changing character of theatrical architecture, the processuality inherent in it, from the birth to the development of the type, and to understand how theatre is an urban happening. Finally, strong links and correspondences are established with the reading of the Palladian project, and in particular with the first and most indicative example of theatre that marked the return and, at the same time, the overcoming of classical concepts, the Olympic Theatre in Vicenza. The second part identifies the primitive actions of foundation, of institution of the theatre building: fencing and covering. Finally, the third and last section explains the synchronic variants, at the base of distinct formation-constructive trends.

Keywords Urban knots · Entertainment facilities · Olympic Theatre · Andrea Palladio

1.1 The Palladian Education by Aldo Rossi

«when he arrived in Venice from the water, my Portuguese friend, José Charters, reminded me that there is a popular proverb or sentence in Portugal that says that all that is good comes from the sea».[1]

Rossi says this about his Teatro del Mondo, an object which, coming from the sea, would seem absolute, most notably detached from the urban context, the expression of a closed and autonomous shape; in its intimate essence, it is instead *open to the world to which it delivers pieces of reality and dreams of unreality.*[2]

Over time the observation, analysis and reconstruction methods of theatre have been various, they have transferred the dynamism of the building, mobile on the water just as mobile and changeable is theatre itself, to interpretation, to criticism, still evolving.

© Springer Nature Switzerland AG 2022
S. Clemente, *Reconstructing Theatre Architecture*, The Urban Book Series,
https://doi.org/10.1007/978-3-030-89968-4_1

One of the most recent implications and a point of arrival in this sense was the exhibition *Aldo Rossi. Theatres* held from June 20 to November 25, 2012, at the Emilio and Annabianca Vedova Foundation in Venice, curated by Germano Celant and staged by Gae Aulenti.

The feature of mobility and more generally of changeability of the theatre building corrodes the fixed aspect of architecture as a simple container of the show. Architecture is continually generated and regenerated, as under the activities of carpenters, bricklayers, electricians, blacksmiths, decorators, set designers, costume designers, who transform the show space while keeping it alive.

In the Teatro del Mondo (Fig. 1.1), a further and highly original aspect of mutability is also created, given by the relativity of relations with space, not only internal but also external. In fact, Gino Malacarne observes that, *being the theatre on the water, you could see the passage of vaporettos and ships from the window as if you were on another ship and these other ships entered the image of the theatre constituting the true fixed and mobile scene.* And again: *the Teatro del Mondo moves towards the Venetian domes and monuments and sails, ideally confronting itself with the Venetian civilization of the Adriatic Sea, up to Dubrovnik. As it passes, it creates new and unexpected urban compositions, inventing possible "analogous cities" with the places it passes through.*[3] The Teatro del Mondo is therefore a multi-level theatrical machine, it is an observation point also conceived to be observed. Inside it is possible to stage the show as well as to stage itself outside, producing the manifestation of many scenes within the same well-known setting, creating ever-changing relationships and thus rediscovering the architecture of the city. It is subject to a double form of verification, in carrying out the functions for which it was designed and during the trip.

Patrizia Montini Zimolo writes: *Aldo Rossi, speaking of Palladio, dwelt on the consideration "that archeology and scenography often meet in his work, in an exemplary way in the Olympic Theatre in Vicenza. [...] Here the interior and exterior are confused, the interior pretends the conditions that are of the city and the city reads the building for what happened inside".*[4] We can therefore rightly say that in the Teatro del Mondo this observation, this lesson of Palladio has been accepted.

Aldo Rossi specifies that this *particular character of Venetian architecture derives from an extraordinary relationship between internal and external spaces that are placed on the same floor. [...] a Venetian interior can never live only from its quality as an interieur but must inevitably belong to the seas of Venice and the woods that plow them, or to become a machine for spying on Venice and its fantasies where interior and exterior once again merge, private and public, individual life and collective life come together as intimately as on* ships.[5]

Theatre thus becomes an exclusively urban event, generated by the city still today as well as at the late medieval origins of the typological process. In summary, in the Teatro del Mondo, it can be said that the changing essence of architecture coincides and is identified with the renewed revelation of the show, no longer only inside but also outside, and therefore urban again.

On the constructive level, the Rossi's theatre is an evident reference to the American Lighthouses, but at the same time recalls the so-called "architectural sculptures",

1.1 The Palladian Education by Aldo Rossi

Fig. 1.1 Aldo Rossi—Teatro del Mondo

typical from the fifteenth century onwards, works by architects such as Brunelleschi, Bernini, Gherardi, which were used to celebrate holidays, wedding ceremonies and funeral events regarding popes and emperors, effectively putting architecture, theatre and scenography into a system.

Most of the times simple structures, certainly temporary, real fragments deriving from the occasions are in fact regarded as the basis of the developing process of the space for the show.

In the following, some statements and definitions by Aldo Rossi himself have been reported to clarify even better the intentions and peculiar characteristics of a project constituting a hub of contemporary theatrical architecture.

Differently, the project for the Teatro del Mondo or for this Venetian theatre is characterized by three facts, having a precise even if unspecified usable space, placing itself as a volume according to the shape of Venetian monuments, being on the water [...].

These analogies of place in designing a building are of decisive importance for me, if properly read they are already the project. Although this is a predictably short-lived building, it is not just a Venetian whim [...].

Finally, theatre, stable or temporary, was a large work of carpentry barely masked by gold and stucco. These are the few notes on one of my projects, independent of the possibility of its construction and its use. But certainly not independent from a Venetian construction, from a way of designing that seeks fantasy only in reality [...].

Venice is a liminal city between water and sea; indeed it is the city par excellence that lives this relationship. An ancient and always precarious relationship where the construction of man and the transformation of nature are the pre-eminent aspect of architecture. Also in the Venice of white stone stopped by the Palladian monuments there is this memory of a port, wooden, mercantile construction; where stone is the material that confirms the first appropriation of the place.[6]

Aldo Rossi, therefore, defines the specific relationship of the project with the Venetian urban context and from this moves towards broader considerations relating to the relationship of ephemeral architecture with the old town. A relationship that, as mentioned so far and as it will be seen a little further on, is inherent in the very birth of the theatre from the city, or rather from the temporary representations in the existing, in the context of the urban fabric.

Tafuri writes that, in the reality of Venice [...] "its evidence [of theatre] ruthlessly tears off masks that time has glued to faces". This last observation, regardless of the theatre I built, is an important starting point for the question of historic centers; I believe that the new buildings will have to rediscover the architecture of the ancient city and will not be able to somehow consider the mask glued to her face. Even if this mask is perhaps the charm from which we cannot detach ourselves and is destroyed and destroyed it must remain; because every intervention is, by its nature, merciless. But these problems—a real, progressive, unusual alternative—were certainly not addressed. It was preferable to understand the ephemeral as a low-cost construction, a shack, a carousel; and also to transform the poetry of the shacks into the more ambitious guise of festivals and the hasty initiatives of the municipal administrations. It is also good that what you don't like and poses too many problems is ephemeral.[7]

In this last sentence, Rossi clarifies the thin boundary of temporariness, which must not be understood as a form of protection towards the real change of our cities, but as a generating force, just as happened in the developing process of modern space for the show, which in this work we are about to analyze.

The city, Rossi always says, is the necessary result of past circumstances. Hence, its transposition into time and memory, which is the only possible way to alleviate the existence of architecture that does not want to be rigidly determined and decisive, but the emanation of a chain of remote causes, with effects that can lead to future.

Drawing and representation in general deserve a separate discussion. These necessarily have an intrinsic relationship with design and research. The medium, the means constituted by drawing, is especially in this case essential for the morphological, urban and historical-architectural understanding of theatre.

And theatre is perhaps the only building capable of fulfilling the function of representation, of identifying the culture of a country and at the same time being a reference not only topographical of the city.

In conclusion, as reported by Vera Rossi and Margherita Guccione: *theatre is therefore eternal and firm, yet constantly changing and ductile, it is located in a peculiar place but creates a universal space. By its nature it is place in place, life in life, beginning and end. Thus architecture is the scene where existence begins, unfolds and ends. In this sense, we believe that for Rossi theatre represented the symbolic synthesis of architecture itself.*

1.2 Fencing and Covering

When something happens in a flat place, everyone rushes, those who are behind try in every way to get up on those who are in front: you get on the benches, you roll the barrels on the spot, you approach with the carriage, boards are brought from all sides, a nearby hill is occupied and a crater is formed so quickly. If the show is repeated in the same place, a stage is erected for those who can pay, and the rest of the people sit down as best they can. The architect's task is to satisfy this general desire.[9]

In this statement, it is clear that theatre should be understood rather as a "theatrical place", born from experience, extremely linked to the events that produce its form.

Luigi Allegri traces two interesting lines of research at the origin of modern theatre through the medieval period. The first is to seek theatricality outside the classical theatre building. *The Middle Ages lost for a long time the memory of the idea of theatre codified in antiquity, and took centuries and very long processes to elaborate and above all to impose new ones, but in the meantime it uses isolated elements among those whose constellation constitutes theatre, to maintain the anthropological function of theatricality in life.*[10] The second strand is reserved for the category of the spectacular. This aspect was among the few not subject to condemnation by the Church, which instead saw in it an important and powerful educational tool to be offered to the public of the faithful, to the people.

In the Middle Ages, the discrepancy existing in Roman times between the monumentality of the theatrical building and the show that took place inside was sharpened. The progressive emptying of the meaning and quality of the representations also contributed to the abandonment of buildings that no longer met the needs, marking a clear break with the past. In fact, in the Middle Ages that very long process that leads to the re-foundation of theatre as building starts, opening up a new typological process.

However, if we stop at the medieval theatre alone, this, it has been said, has not produced in the history of Italian and European architecture the construction of places suitable for representation, conceived, designed and built with a single and specific purpose. When theatre as a structured activity no longer exists, the anthropological demands and the technical skills that are peculiar to it are disseminated in a widespread theatricality that is inserted in other moments and in other places of social life. Above all, widespread theatricality finds a fruitful context on festive occasions, both community occasions, which involve the entire city, and the more particular and private ones as weddings or banquets. Drama finds its natural setting in the spaces of the community, in the streets, in the squares. These are multifunctional areas par excellence, considered able of temporarily hosting theatre, the exceptional nature of the occasion; in the moment of the performance the square becomes an image of the theatre, the city and the cosmos.

Grafting on the different urban aggregations, on the pre-existing fabrics, the scenic needs, the "functions" of the representation orient their creative reuse.

The primary form of the constructive act of fencing is therefore identified in the action of recognition of the pre-existing place. The nodal axis coincides with the main entrance to the space, initially outdoor, used to host the show. *The nodal axis identifies a central place that unifies the structure and use of space in a single constructive gesture*[11] to which symbolic value is recognized. The direction of the axis establishes the succession of the scenic elements, determines their hierarchy, guides their reading.

The node is instead the nucleus of the closed space, intersection and center of the paths, sometimes indicated by symbolic elements such as a tree, the water well, etc., focal points.

The "closure", the peripheral boundaries of the square effectively cancel the distinctions that were originally evident between courtyards and squares. As happened for the cloister of the Franciscans in Romans sur Isère, an old private courtyard has become a public place.

In an ideal expansion from the center towards the more peripheral areas of the building, the city can be seen as a circular pattern crossed by its directional axes. In the real dimension, the city center often does not coincide with the geometric center of the city; the narrow, cluttered and winding medieval street, generated by the urban fabric developing process, by the attraction towards the church, the castle or the market, is not ideal. The ideal cities of the urban planners of the Renaissance will rediscover the ancient practices of the Romans, to trace the project of the city, to mark coinciding real and ideal centers. In the representation of medieval drama, it is the scenographic system that is placed in a multiform and multifunctional constructed context, which imposes certain constraints. However, the influence is reciprocal, opening theatre to different ways of interpreting and using the existing. The correspondence between real and ideal should be understood as a more or less exact approximation that you are trying to reach.

So far, the square has been identified as the boundary of the enclosure, and yet it is not the only perimeter involved in the theatrical representation. The disposition of the public on one side and the delimitation of the stage, of the area intended for the realization of the real drama, are in turn other means and ways of declining, of personalizing the city public space. However, this organization does not create any internal fracture, on the contrary it rejects the break between the space of representation and the spectator. There is a sharing in some ways very similar to the essence of the early Greek theatre.

Religious theatre, understood as a whole, is the most macroscopically documented phenomenon of medieval theatricality. However, paradoxically, contemporary theorists missed the theatrical aspects of the religious dramas themselves, to condemn the show was a base of their critical thinking. In this regard, the in-depth analysis by Luigi Allegri is reported below, focusing in particular on the writings of John of Salisbury and Hugues de Saint-Victor.

Now we should be satisfied to believe that for John of Salisbury (Policraticus just beyond the mid-twelfth century), the idea of theatre is not only present but also well defined. A further proof is in his insistence on the metaphor of the theatre of the world—a metaphor that indeed owes a renewed fortune to him after centuries of obsolescence -, handling which, he demonstrates perfectly possessing the mechanisms of theatrical representation: "the life of man on earth is a comedy in which each, forgetful of himself, plays the part of another […]. Thus, to adapt the inventions of the Gentiles to the pious ears, it will be said that the end of all things is tragic. But I have nothing to object to if we want to keep the term "comedy" as more pleasant; since it is also known among us that—as Petronius says—almost everyone on earth behaves like histrions. "[…].

Let's go in search of confirmations, approaching the other example, that of Hugues de Saint-Victor. He too has dealt with theatre and indeed, as we have seen, in the Didascalicon, he introduced theatrica among the mechanical arts, thus institutionalizing and legitimizing it as an anthropologically central activity. Well, even Hugues places theatre and all that concerns it in the past and does not mention contemporary theatricality: neither, it was said, that of jesters, nor, let us add now, of religious. And if an attempt was made to identify the reasons for the first censorship in the intent consistent with the culture of the time to marginalize jesterery and its competitive demands to those of the Church, for the second absence this justification naturally does not apply. And therefore all that remains is to conclude that even for Hugues, what for us is religious theatre, for him is certainly something religious but not theatre […].

Hugues de Saint-Victor reconstructs the reasons for its construction in the ancient world: to prevent the show from being made in other places, and since the show was unavoidable from society, buildings specifically used for this were needed. It should therefore be natural that, once it is recognized that shows exist and cannot be eliminated even in medieval culture, new theatrical buildings should be built, especially if we consider that it is precisely the intrinsic danger of the show that recommended, according to Hugues, the identification of places for its ghettoization. If this has not happened, it cannot be left to pass over in silence, without problematization. And here, again, we probably clash with that unresolved ambiguity that is of medieval Christian culture towards theatre but also of modern historiography on a contradictory phenomenon [...].

Given the premises, the Christian Middle Ages cannot, literally, legitimize theatrical practice with an act that would result in an unambiguous declaration, such as the construction of theatres. If theatricality is ideologically condemned, it is necessary that when one accepts it or undergoes it practically, it is not located in specific spaces—which would institutionalize it—but in spaces already given and whose symbologies are not censurable, such as the church, the oratory or the square. Only in this way theatricality can be felt as an ephemeral activity, substantially extraneous to the social fabric, just as the space that hosts it is a foreign body that temporarily overlaps the urban fabric without leaving architectural residues. Thus theatre qualifies as a practice that does not take place and that has no name, which has no definition in the defining universe of the encyclopedias of medieval knowledge, and which therefore cannot exist institutionally [...].

And these are only two examples of this glaring silence of theory, in the face of which the few testimonies of stance, the ones that scholars struggle to report every time because they are the only ones to offer us fragments of documentation, really appear like small islets in the middle of the ocean. But this is only one of the terms of the speech; the other focuses precisely on the analysis of these surviving testimonies of awareness of the theatrical characteristics of religious dramas. There are documents that recognize these characteristics, and also claim their positivity, especially in comparison with the paganistic theatricality of the festivals of the fools, such as the gloss to Innocent III and the decree of Alfonso X [...], or a canon of the provincial council of Sens of 1460. Or texts that even more clearly seem to qualify these ceremonies as theatre tout court, such as the Horatian commentary between the ninth and eleventh centuries [...], or the much-cited chronicle of the sacred drama performed in Riga in 1204, in which there is an explicit identification of this type of show with the Latin comedy [...].

If the model described here is correct, the path would then be the opposite of what has traditionally been maintained: not from the liturgical, due to progressive extensions, to the theatrical, but from the theatrical, due to progressive neutralization of the dramatic, to the liturgical.

[...] the drama in Latin gives the drama in the vernacular [...] also an idea of stage space that is absolutely original in the history of the show, with that fragmentation of the places of action, the sedes—or loci or domus or mansiones—which force the viewer into a relativized and not absolute relationship that will happen later with the space of the show.[12]

The two plans by Cysat [(Fig. 1.2)] for the representation of Lucerne (Evans, 1943) not only depict a square whose constituent elements are integrated into a

1.2 Fencing and Covering

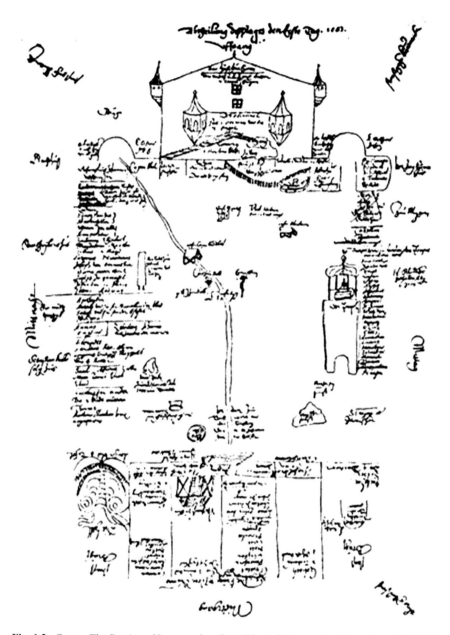

Fig. 1.2 *Cysat—The Passion of Lucern, plan,* [http://theaterhistoryonline.blogspot.com/2014/08/lucerne-passion-play.html]

theatrical place, but also show the ideal and mythical world from which Christian civilization was ordered, integrated into the urban structure. In the same way that medieval cartography placed Jerusalem at the center of the world, and faith placed Passion at the center of human history, the scenic topography established in the living heart of the city—the market square—the cartographic reduction of sacred space. [...] The map of the world of Passion extends to the same level as the ground, with its rivers, its mountains, its cities. If all the places of action are present simultaneously on the scene, it is really the scene of the world. In every moment of drama, the viewer is faced with the totality of this world delimited by Hell or Heaven. [...] An urban scenography is therefore not necessary, since the scene is the real, palpable urban element, the one in which the citizen finds the social center and the forces that govern his universe gathered together.[13]

Thus, also the Villingen Passion constitutes a process of spatial development in the built and from the built environment towards an organic definition of theatre. From interpretation, from reading the space to a new space.

1.3 Synchronic Variants: Central Stage and Circular Theatre

If the first act of construction is fencing, from this, we move on to the increasingly complex articulation of the central scene and the search for autonomy from the context, distinguishing it clearly. The scene is a space unto itself, as the city and spectators are. The orientation and the center of the scene coincide with the ideal center of the city and the cosmos. Similar examples can be found in the Nativity of Rouen or in the more famous Passion of Valenciennes, in which there is respectively the frontal arrangement on the square and frontal on the stage.

The square [...] however remains a fundamental model from which these theatres are organized, and after all the scenic arrangement of the single stage is just a variant (we could say synchronic)—when it is in a central position—of the theatrical square. However, the spatial rupture introduced by the stage highlights that theatre is no longer experienced in the same way and that the show, with all that this term implies as a limitation of ceremonial participation, prevails.[14]

Another synchronic variant of the stage can be considered the circular theatre. Undoubtedly, the result of a more marked abstraction, of a lesser conditioning given by the surroundings. In this regard, see *The Castle of Perseverance*. Richard Southern argued that the performance area was built in this case as a natural theatre. The outer ring was filled with water, a real moat or made up of barriers to hinder non-paying spectators while the paying public was sitting on a hill built right inside the ring. Spectators were free to sit down, to stay near the center or to move around the area intended for the show. Southern estimated that the theatre in question could be about 30 m in diameter and could accommodate an audience of a thousand or more people. There were also boxes located around the ring, divided into separate areas.

1.3 Synchronic Variants: Central Stage and Circular Theatre

These would have been connected to the center by passages through the crowd. This representation could be configured, unlike the others analyzed so far, as a traveling show, which was represented by the same traveling company. The place was therefore recreated every time, adapted, released, abstracted precisely from the context in which it was placed from time to time.

See also, with regard to the circular theatre, another milestone, a figurative source par excellence of the medieval show, a pictorial work that summarizes the representation: *The martyrdom of Saint Apollonia* by Jean Fouquet, painted in Tours between 1452 and 1461. The executioners rage with extravagant gestures on the body of the Saint. All around a crowd gathered in a semicircle on a grandstand witnesses the torture of the extraction of the teeth. Apollonia quietly endures, her facial features fixed in a mask of bliss. A man reads a book and seems to lead the scene, the Roman Emperor Decius with his courtiers seems to want to persuade the Saint to abjure, a fool takes off his shoes and uncovers his buttocks as a sign of contempt. On the grandstand, among the undisciplined public, various places of action are identified. In the center, the empty throne of the Emperor. On the left, Heaven, God dressed in red and St. Michael dressed in gold, on the right, Hell, whose jaws open to show the damned, some characters appear and disappear among the clouds of smoke. Men covered in cloths and armed with clubs, devils and angels complete the scene.

The work measures just 16×12 cm, a miniature, on which Fouquet worked with a magnifying glass and black glass. The Martyrdom was one of the 47 miniatures that literally illuminated a Book of Hours, commissioned by Etienne Chevalier, intended for the laity to strengthen their faith, accompanying them in prayer throughout the seven hours of the day according to the rhythm of the canons. With Fouquet, the miniature becomes a real painting, and the individual miniatures have been cut out and transformed into independent works. Fouquet does not use the typical devotional figures, instead he invents completely original compositions. Elements of realism in the martyrdom of St. Apollonia are the wooden boxes that delimit a medieval theatrical scene. The man with the baton is an arbiter, angels, devils and smoke are the secrets (today's special effects), a real mystery therefore. The work is a unique testimony, since the mysteries by their nature are ephemeral, destroyed at the end of each representation. Through the filter of theatre Fouquet finally achieves a form of detachment, albeit minimal from the violence expressed.

The model of the Elizabethan elastic-wooden theatre derives from this variant, which does not find fertile ground in Italy, but is influenced by the plastic theatre of proper Italian development.

The action of fencing is followed by that of covering. The audience, in front of the single stage, comes to gather in a special space, in a sort of open-air room, covered by a *velum*, with seats, tiers and boxes to represent a new social hierarchy.

Closing also means going to emphasize the growing importance of vertical axiality. The stage has vulgarly an above and a below, prefigurations of what will be trellises and movable bridges in the modern theatre.

Equivalently on the horizontal plane, the back wall of the scene appears, probably borrowed from the tableau vivant, to delimit the space of the representation. With the rear edge of the scene, there is also the space subtracted from the scene, the rear stage.

Here, therefore, is the theatrical organism in germ as a set of elements linked together by a relationship of necessity, competing towards the same purpose of representation.

1.4 Summary

We can say that we have underlined, through the analysis of Aldo Rossi's Teatro del Mondo, the changing nature of architecture, the inherent processuality, as well as having identified the relationships between the exterior and interior of the building as fundamental. Theatre is an urban event as it is generated, produced, by the urban fabric.

Attention was then brought back, through the analysis of case studies, to the primary construction actions, characterized by a minimum level of typicality, belonging to each architectural organism, of fencing and covering.

The simplest form is that of the enclosure, of the square area or more generally of the city, which is identified as suitable for fulfilling the special, temporary theatrical function.

Essentially two synchronic variants were then recognized, relating on the one hand to the independence of the central stage, and on the other to the creation of circular theatres, abstract models, often linked to travel, companies, forms therefore distant from alternative "site-specific " installations.

1.4 Summary

IL TEATRO DEL MONDO   CODE1.01

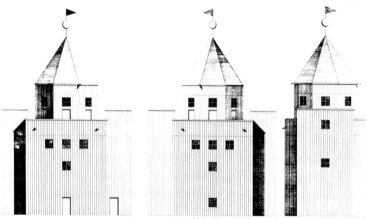

Aldo Rossi - Il teatro del mondo - 1979

TYPOLOGICAL SCHEME

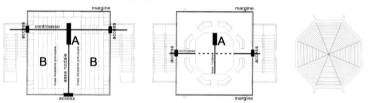

plan of the ground floor, plan of the fourth floor, plan of the roof

LEGEND

A SCENE   B AUDIENCE

SINTESI DEI PRINCIPALI CARATTERI DI SELEZIONE

luogo_**Venezia**
livello della manifestazione_**internazionale**
sito di interesse architettonico o storico/artistico/archeologico_**non più esistente**
sito UNESCO_**Si**
anno_**1979**
realizzazione_**teatro temporaneo**
materiale costruttivo prevalente_**acciaio e legno**
caratteristiche_**scena distinta dall'area dedicata agli spettatori**
processo_parola chiave_**recinto**

THE PASSION OF VILLINGEN — CODE1.02

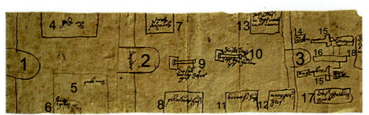

The passion of Villingen - 15th century

TYPOLOGICAL SCHEME

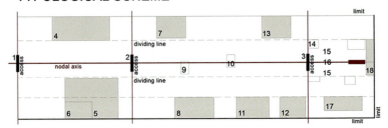

LEGEND

1 first door
2 second door
3 third door
4 Inferno
5 Garden of Gethsemane
6 Mount of Olives
7 Herod
8 Pilate
9 column of the Flagellation
10 column of the Rooster
11 Caiaphas
12 Anna
13 Last Supper
14 Tombs
15 crosses of the robbers
16 cross of Christ
17 Holy Sepulcher
18 the Sky

SUMMARY OF THE MAIN SELECTION CHARACTERS

place_**Villingen (GERMANY)**
level of the event_**local**
site of architectural interest_**demolished**
UNESCO site_**No**
year_**15th century**
building_**temporary theatre**
construtcion_**wooden**
features_**multiple scene not distinct from the area dedicated to spectators**
keyword_**enclosure**

1.4 Summary

THE PASSION OF VALENCIENNES CODE1.03

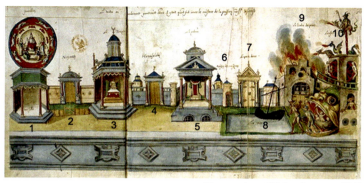

The passion of Valenciennes - 1547

TYPOLOGICAL SCHEME

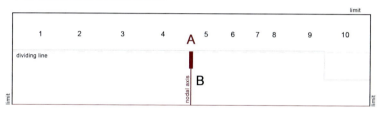

LEGEND

A SCENE   3 the Temple   6 the house of the bishops   9 thw limbo of the fathers
1 Paradise   4 Jerusalem   7 the golden door   10 Inferno
2 Nazareth   5 the palace   8 the sea   B AUDIENCE

SUMMARY OF THE MAIN SELECTION CHARACTERS

place_**Valenciennes (FRANCE)**
level of the event_**local**
site of architectural interest_**demolished**
UNESCO site_**No**
year_**1547**
building_**temporary theatre**
construction_**wooden**
features_**multiple scene distinct from the area dedicated to spectators**
keyword_**enclosure**

THE CASTLE OF PERSEVERANCE — CODE1.04

Inside the circle:
This is the Castle of Perseverance that stands in the midst of the place, but let no men sit there for letting of sight [obstructing the view], for there shall be the best [position] of all. [...]
Between the circles:
This is the water about the place if any ditch may be made where it shall be played, or else [see] that it be strongly barred all about; and let not over many stytelerys be within the place.

The Castle of Perseverance - 1461

TYPOLOGICAL SCHEME

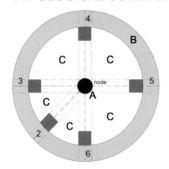

LEGEND

A MAIN PLACE OF ACTION
1 The Castle

B MOAT
C AUDIENCE

2 North-East Coveytyse Scaffold
3 East God's Scaffold
4 South Caro
5 West Mundus Scaffold
6 North Belial Scaffold

SUMMARY OF THE MAIN SELECTION CHARACTERS

place_**itinerant**
level of the event_**local**
year_**1461**
building_**temporary theatre**
construction_**wooden**
features_**multiple scene distinct from the area dedicated to spectators**
keyword_**enclosure, node**

1.4 Summary

THE MARTYRS OF ST. APOLLONIA   CODE1.05

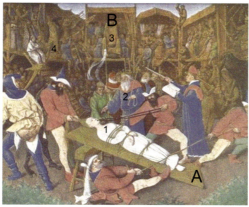

Jean Fouquet – The martyrs of St Apollonia - 1461

TYPOLOGICAL SCHEME

LEGEND

A MAIN PLACE OF ACTION
1 St Apollonia
2 group of the Emperor Decius

B SCENE
3 imperial throne
4 Paradise
5 Inferno

C AUDIENCE

COMPARATIVE TABLE — CODE 1.26

MULTIPLE SCENE NOT DISTINCT FROM THE AREA DEDICATED TO SPECTATORS

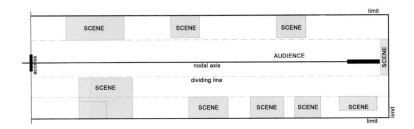

SYNCHRONIC VARIANTS

MULTIPLE SCENE DISTINCT FROM THE AREA DEDICATED TO SPECTATORS

CENTRAL STAGE

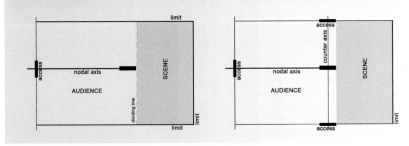

CIRCULAR THEATRE

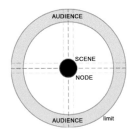 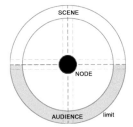

Notes

1. «quando giunse a Venezia dall'acqua, il mio amico portoghese, José Charters, mi ricordò che esiste in Portogallo un proverbio o sentenza popolare che dice che tutto ciò che è buono viene dal mare». Rossi A (1982) Teatro del Mondo. CLUVA, Venezia: 11–16.

2. *«aperto al mondo a cui consegna brani di realtà e sogni di irrealtà».* Celant G (ed) (2012) Aldo Rossi Teatri. Skira, Milano: 7.

3. *«stando il teatro sull'acqua si vedeva dalla finestra il passaggio dei vaporetti e delle navi come si fosse su un'altra nave e queste altre navi entravano nell'immagine del teatro costituendo la vera scena fissa e mobile. E ancora: il Teatro del Mondo si muove verso le cupole e i monumenti veneziani e naviga, confrontandosi idealmente con la civiltà veneziana del mare Adriatico, fino a Dubrovnik. Al suo passaggio crea nuove e impreviste composizioni urbane, inventando con i luoghi attraversati, possibili "città analoghe"».* Malacarne G, Montini Zimolo P (2002) Aldo Rossi e Venezia. Il teatro e la città. Edizioni Unicopli, Milano: 72.

4. *«Aldo Rossi parlando del Palladio si era soffermato sulla considerazione "che archeologia e scenografia si incontrano spesso nella sua opera, in modo esemplare nel Teatro Olimpico di Vicenza. [...] qui interno ed esterno si confondono, l'interno finge condizioni che sono della città e la città legge l'edificio per quel che è avvenuto al suo interno"».* Ibidem: 43.

5. *«particolare carattere dell'architettura veneta deriva da uno straordinario rapporto tra spazi interni ed esterni che vengono posti su uno stesso piano. [...] un interno veneziano non può mai vivere soltanto della sua qualità di interieur ma deve fatalmente appartenere ai mari di Venezia e ai legni che li solcano ossia diventare una macchina per spiare Venezia e le sue fantasie dove interno ed esterno ancora una volta si confondono, privato e pubblico, vita individuale e vita collettiva si uniscono intimamente come sulle navi».* Rossi A (1982) Teatro del Mondo. CLUVA, Venezia: 11–16.

6. *«Differentemente il progetto per il Teatro del Mondo o chiamiamolo per questo teatro veneziano si caratterizza da tre fatti, l'avere uno spazio usabile preciso anche se non precisato, il collocarsi come volume secondo la forma dei monumenti veneziani, l'essere sull'acqua [...].*

   Queste analogie del luogo nel progettare un edificio hanno per me un'importanza decisiva, se ben lette sono già il progetto. Anche se si tratta di un edificio dal tempo prevedibilmente breve esso non è solo un capriccio veneziano [...].

   Infine il teatro, stabile o provvisorio era una grossa opera di carpenteria appena mascherata dagli ori e dagli stucchi. Queste sono le poche note su un mio progetto indipendenti dalla possibilità della sua costruzione e dal suo uso. Ma certamente non indipendenti da una costruzione veneziana, da un modo di progettare che cerca solo nel reale la fantasia [...].

   Venezia è una città liminare tra acqua e mare; anzi è la città per eccellenza che vive questo rapporto. Rapporto antico e sempre precario dove la costruzione dell'uomo e la trasformazione della natura costituisce l'aspetto preminente dell'architettura. Anche nella Venezia di bianca pietra fermata dai monumenti palladiani vi è questa memoria di una costruzione portuale, lignea, mercantile; dove la pietra è il materiale che conferma la prima appropriazione del luogo». Ibidem.

7. *Tafuri scrive che, nella realtà di Venezia, "... la sua evidenza (del teatro) strappa impietosamente maschere che il tempo ha incollato ai volti". Quest'ultima osservazione, indipendentemente dal teatro da me costruito, è uno spunto importante per la questione dei centri storici; credo che le nuove costruzioni dovranno riscoprire l'architettura della città antica e non potranno in qualche modo considerare la maschera incollata al suo volto. Anche se questa maschera costituisce forse il fascino da cui non riusciamo a staccarci ed è distrutto e distrutto deve rimanere; perché ogni intervento è, per sua natura, impietoso. Ma questi problemi—un'alternativa reale, progressiva, insolita— non erano certo raccolti. Era preferibile intendere l'effimero come costruzione da pochi soldi, una baracca, una giostra; e anche tramutare la poesia delle baracche sotto le vesti più ambiziose dei festivals e delle frettolose iniziative delle amministrazioni comunali. E' bene anche che sia effimero ciò che non piace e pone troppi problemi.* Ibidem.

8. *«il teatro è quindi eterno e fermo, eppure costantemente mutevole e duttile, si trova in un luogo peculiare ma crea uno spazio universale. Per sua natura è luogo nel luogo, vita nella vita, inizio e fine. Così l'architettura è la scena dove inizia, si svolge e termina l'esistenza. In questo senso crediamo che per Rossi il teatro rappresentasse la sintesi simbolica dell'architettura stessa».* Celant G (2012) Aldo Rossi Teatri. Skira, Milano: 8.

9. *«Quando, in un luogo piano accade qualche cosa, tutti accorrono, quelli che sono indietro cercano in tutti i modi, di sollevarsi su quelli che sono innanzi: si sale sui banchi, si fanno rotolare le botti sul posto, ci si avvicina con la carrozza, si apportano tavole da ogni parte, si occupa un'altura vicina e si forma così in fretta un cratere. Se lo spettacolo si ripete al medesimo posto, si erige un palco per quelli che possono pagare, e il resto della gente si accomoda come può. Il compito dell'architetto è quello di soddisfare a tale desiderio generale».* Goethe JW (1991) Viaggio in Italia (1786–1788). Rizzoli, Milano.

10. *«Il Medioevo perde per molto tempo la memoria dell'idea di teatro codificata nell'antichità, e impiega secoli e processi lunghissimi ad elaborarne e soprattutto ad imporne di nuove, ma nel frattempo utilizza elementi isolati tra quelli la cui costellazione costituisce il teatro, a mantenere in vita appunto la funzione antropologica della teatralità».* Allegri L (1988) Teatro e spettacolo nel medioevo. Editori Laterza, Roma: IX.

11. *«L'asse nodale individua un luogo geometrico accentrante che unifica struttura e uso dello spazio in un unico gesto costruttivo».* Strappa G (1995) Unità dell'organismo architettonico. Edizioni Dedalo, Bari: 82.

12. *«Ora accontentiamoci di ritenere come per Giovanni di Salisbury (Policraticus poco oltre la metà del XII secolo), l'idea di teatro non sia solo presente ma anche ben definita. Una riprova ulteriore è nel suo insistere sulla metafora del teatro del mondo—metafora che anzi proprio a lui deve una rinnovata fortuna dopo secoli di obsolescenza -, maneggiando la quale dimostra di possedere perfettamente i meccanismi della rappresentazione teatrale: "la vita dell'uomo sulla terra è una commedia in cui ciascuno, dimentico di sé, recita la parte di un altro [...]. Così, per adattare alle pie orecchie le invenzioni dei gentili,*

1.4 Summary

si dirà che la fine di tutte le cose è tragica. Ma non ho nulla da obiettare se si vorrà mantenere, come più gradevole, il termine "commedia"; poiché è risaputo anche fra di noi che—come dice Petronio—quasi tutti, sulla terra si comportano da istrioni." […].

Andiamo in cerca di conferme, accostandoci all'altro esempio, quello di Ugo di San Vittore. Anch'egli si è occupato di teatro e anzi, come abbiamo visto, nel Didascalicon, ha introdotto la theatrica tra le arti meccaniche, dunque istituzionalizzandola e legittimandola come attività antropologicamente centrale. Ebbene, anche Ugo colloca nel passato il teatro e tutto quanto gli attiene e non fa parola della teatralità contemporanea: né, si diceva, di quella dei giullari, né, aggiungiamo ora, di quella religiosa. E se della prima censura si è tentato di individuare le ragioni nell'intento coerente con la cultura del tempo di marginalizzare la giulleria e le sue istanze concorrenziali a quelle della Chiesa, per la seconda assenza questa giustificazione naturalmente non vale. E dunque non resta che concludere che anche per Ugo, ciò che per noi è teatro religioso, per lui certo è qualcosa di religioso ma non teatro […].

Ugo di San Vittore ricostruisce le ragioni della sua costruzione nel mondo antico: per impedire che si facesse spettacolo in altri luoghi, e dato che lo spettacolo era ineliminabile dalla società, furono necessari edifici specificamente adibiti a ciò. Dovrebbe dunque essere naturale che, una volta riconosciuto che gli spettacoli esistono e sono ineliminabili anche nella cultura medievale, si procedesse alla costruzione di nuovi edifici teatrali, soprattutto se si considera che è proprio l'intrinseca pericolosità dello spettacolo che ha consigliato, secondo Ugo, l'individuazione di luoghi per una sua ghettizzazione. Se questo non è avvenuto, non è dato da lasciare passare sotto silenzio, senza problematizzazione. E qui, ancora, ci si scontra probabilmente con quella ambiguità irrisolta che è della cultura cristiana medievale nei confronti del teatro ma poi anche della storiografia moderna su un fenomeno contraddittorio […].

Date le premesse, il Medioevo cristiano non può, letteralmente, legittimare la pratica teatrale con un atto che risulterebbe una dichiarazione senza ambiguità, come la costruzione di teatri. Se la teatralità è condannata ideologicamente, è necessario che quando la si accetti o la si subisca praticamente, non la si situi in spazi specifici—che la istituzionalizerebbero—ma in spazi già dati e le cui simbologie non sono censurabili, come la chiesa, l'oratorio o la piazza. Solo così la teatralità può essere sentita come attività effimera, sostanzialmente estranea al tessuto sociale così come lo spazio che la ospita è un corpo estraneo che si sovrappone temporaneamente al tessuto urbano senza lasciare residui architettonici. Così il teatro si qualifica come una pratica che non ha luogo e che non ha nome, che non ha una definizione nell'universo definitorio delle enciclopedie del sapere medievale, e che dunque non può esistere istituzionalmente […].

E non si tratta che di due soli esempi di questo vistoso silenzio della teoria, di fronte al quale le poche testimonianze di presa di posizione, quelle che gli studiosi si affannano ogni volta a riportare perché sono le sole ad offrirci frammenti di documentazione, figurano davvero come isolotti sparuti in mezzo all'oceano. Ma questo è solo uno dei termini del discorso; l'altro verte proprio sull'analisi di queste testimonianze superstiti di consapevolezza delle caratteristiche teatrali dei drammi religiosi. Ci sono sì i documenti che riconoscono queste caratteristiche, e anche ne rivendicano la positività, specie a confronto con la teatralità paganeggiante delle feste dei folli e simili, come la glossa a Innocenzo III e il decreto di Alfonso X [...], oppure un canone del concilio provinciale di Sens del 1460. Oppure testi che ancor più nettamente sembrano qualificare come teatro tout court queste cerimonie, come il commento oraziano tra IX e XI secolo [...], oppure la citatissima cronaca del dramma sacro rappresentato a Riga nel 1204, in cui c'è un'esplicita identificazione di questo tipo di spettacoli con la commedia latina [...].

Se il modello qui descritto è esatto, il percorso sarebbe allora il contrario di quello che si è tradizionalmente sostenuto: non dal liturgico, per via di progressivi ampliamenti, al teatrale, ma dal teatrale, per via di progressive neutralizzazioni del drammatico, al liturgico.

[...] il dramma in latino consegna al dramma in volgare [...] anche un'idea di spazio scenico che è assolutamente originale nella storia dello spettacolo, con quella frammentazione dei luoghi di azione, le sedes—o loci o domus o mansiones—che costringono lo spettatore ad un rapporto relativizzato e non assoluto come avverrà in seguito con lo spazio dello spettacolo». Allegri L (1988) Teatro e spettacolo nel medioevo. Editori Laterza, Roma: 133–135, 174, 230–231.

13. *«Le due piante di Cysat per la rappresentazione di Lucerna (Evans, 1943) non raffigurano solamente una piazza i cui elementi costituitivi sono integrati a un luogo teatrale, ma mostrano ugualmente il mondo ideale e mitico a partire dal quale si è ordinata la civilizzazione cristiana, integrata alla struttura urbana. Nello stesso modo in cui la cartografia medievale collocava Gerusalemme al centro del mondo, e la fede situava la Passione al centro della storia umana, la topografia scenica stabiliva nel cuore vivente della città— la piazza del mercato—la riduzione cartografica dello spazio sacro. [...] La carta del mondo della Passione si estende allo stesso livello del suolo, con i suoi fiumi, le sue montagne, le sue città. Se tutti i luoghi dell'azione sono presenti simultaneamente sulla scena, essa è realmente la scena del mondo. In ogni momento del dramma lo spettatore si trova di fronte alla totalità di questo mondo delimitato dall'Inferno o dal Paradiso. [...] Non si rende dunque necessaria una scenografia urbana, poiché la scena è l'elemento urbano reale, palpabile, quello nel quale il cittadino trova riuniti il centro sociale e le forze che reggono il suo universo».* Konigson E (1990) Lo spazio del teatro nel medioevo. La Casa Usher, Lucca: 51–52.

14. *«La piazza [...] resta comunque un modello fondamentale a partire dal quale si organizzano questi teatri, e in fondo la disposizione scenica del palco unico non è che una variante—allorché si trova in posizione centrale—della piazza teatralizzata. Tuttavia la rottura spaziale introdotta dal palco evidenzia che il teatro non è più vissuto nello stesso modo e che lo spettacolo, con tutto ciò*

1.4 Summary

che questo termine implica come limitazione della partecipazione cerimoniale, prevale». Ibidem: 5.

References

Brusatin M (1979) Venezia e lo spazio scenico. Edizioni "La Biennale" di Venezia, Venezia

Celant G (2012) Aldo Rossi. Teatri. Skira, Milano

Malacarne G, Montini Zimoli P (2002) Aldo Rossi e Venezia. Il teatro e la città. Edizioni Unicopli, Milano

Rossi A (1996) Un'educazione palladiana (Vicenza, Teatro Olimpico, 18 settembre1996). In: Annali di Architettura, Rivista del Centro Internazionale di Studi di Architettura Andrea Palladio (13):9–13

Strappa G (1995) Unità dell'organismo architettonico. Edizioni Dedalo, Bari

Chapter 2
The Theatrical Organism

The Autonomy of the Space for the Show

Abstract The chapter aims to analyze the emergence of the characteristics of autonomy, of recognizability of the space intended for entertainment in Italy, starting with the case study of the Municipal Palace of Ferrara. At the Este court, classical texts, mostly Latin, were represented for the first time, and the first examples of courtyards, later covered, set up for special temporary use were recorded. We therefore want to deepen the descriptive sources, figurative and otherwise, in order to produce hypotheses for the reconstruction of those places, proposing the evolutionary process starting from the courtyard of the house, of the building, from the urban square.

Keywords Theatrical organism · Courtyard · Este court · Ferrara

2.1 Temporarines and Permanency of the Features of the Theatre Space

In medieval theatre, it has been said, the scenic constructions, however grandiose and articulated, are superfetations of architecture and urban areas, they are temporary places that build their own meaning above and in the interstices of the codified spaces of the city. Coherently, the problem of the theatrical building arises from a culture that is free from these ideological presuppositions, such as that of the Renaissance, when even theatre itself is recognized, pursued and established without ambiguity as a cultural practice that is not only legitimate but dignifying and necessary […].

The recovery of an ancient model, historically false but credible and in good faith starting from the data available, can then be configured in reality more as a projection forward, a move to be played in the present, in which a need for theatre is answered, no longer postponable than an attempt at philological reconstruction.[1]

The rediscovery of classical texts constituted a unique engine of development for the so-called Italian scene, then exported to Europe, where until the seventeenth century, the stylistic features of the religious matrix remain, with a consequent deferral of the identification of spaces and the construction of autonomous buildings, exclusively designed for representation. Only a few decades separate the performances mentioned above, spread throughout Europe as well as in Italy, from the shows at the Este court which we will discuss below.

© Springer Nature Switzerland AG 2022
S. Clemente, *Reconstructing Theatre Architecture*, The Urban Book Series,
https://doi.org/10.1007/978-3-030-89968-4_2

A dì 25, la zobia, in la festa de San Paulo. Se recitò la prima comedia de Plauto c.jSiv de Amphitrione e Alchmena in lo cortile novo de la Corte suxo lo tribunale suprascripto, dove he sta' recita' l'altra a dì 21 del prexente, e durò da la prima hora de nocte fino ad hore 6, con sexanta lumere imprexe e altri dupieri; ma non fu finita, perchè il venne una grande piova, la quale cazò le persone, avenga ch'el cortile fosse quasi tutto coperto de tele. E fra li acti forno facte alchune feste; e maxime che l'hera construito uno celo alto a uno cantone verso la torre de l'arlogio con lampade che ardevano a li lochi debiti de drio de tele negre subtile e radiaveno in modo de stelle; e gè herano fanzuli picoli vestiti de biancho in forma de li pianeti, che hera una cosa mirabile da vedere per la grandissima spexa, il quale celo operò a tempo per quello hera necessario per la comedia, con commendatione de tuti li homini intelligenti.

A dì 5, il luni, la festa de Sancta Agata. La comedia de Plauto, la quale non fu finita el dì de San Paulo, che fu a dì 25 del passato, hozi lo illustrissimo duca nostro la fece representar in lo dicto cortile novo: la quale durò da l'hore 21 insino ad hore 3 de nocte, e fu aperto il celo, constructo come fu notato de sopra. Nel quale se sentì cantare e sonare suavemente da cantori perfectissimi, e feceno venire Jove da celo. E finita la comedia, veneno tute le forteze de Hercule per suxo il tribunale predicto, zoè Anteo, le Colonne de Hercule, el Tauro, le Amazone, Centauro, lo Apro, Idra, Cacho con le vache tirate a retro per la coda, e altre molte. E gè hera prexente lo illustrissimo signore marchexe de Mantoa e la illustrissima duchessa nostra, con molti altri signori e zintilhomini e done e gran populo. La spexa fu extimata esser de ducati 1000.[2]

(A. Zamboni).

This excerpt from the Diary of Ferrara describes one of the first representations of the rediscovered Plautian operas in a different context from that of the classical theatre, for which they were conceived[3]: *Amphitryon* which, interrupted by rain, was resumed a few days later. In this and in the subsequent periods, it is clear that the courtyard of the current Municipal Palace was the site of the representation and that it was covered, although with temporary structures, for the occasion given by the representation. Here, the primitive actions of building, enclosing and covering, are represented and testify how the developing process of the modern theatre descended from the transformations of the urban fabric, first temporarily readapted and then definitively transformed with the birth of the autonomous building. In the in-depth information sheet, a reconstructive scheme of the use of the courtyard is proposed, in which the typical aspects of the contemporary religious representation remain, as also described at the European level. The distinction between the area intended for the stage and that occupied by spectators is highlighted, with the very special position of Duke Ercole I on the stage. The ingenuity of Paradise is located at the opposite vertex, the audience arranged in front of the stage on steps. We can assume that the spaces between the steps themselves and the stage were also occupied, as well as the areas below the loggias and arcades. The courtyard becomes, although temporarily, a knot.

2.1 Temporarines and Permanency of the Features of the Theatre Space

In the following lines, we analyze the main transformations of the spaces for the show in the complex of Estense Ducal Palace linked to the construction phases of the same over time.

The context in which the described representation of *Amphitryon* took place is that of a profound reworking of the pre-existing situation up to 1479. From that date and up to about 1481, the Duke Ercole I made the palace assume the almost current conformation, with the courtyard, the garden of the Duchesses and the access staircase by Pietro Benvenuti,[4] as part of an urban development that affected the entire northern area of the city, the so-called Herculean addition by Biagio Rossetti and Pellegrino Prisciani. The building was originally configured, in the middle of the thirteenth century, with Azzo and Obizzo D'Este, with a single floor and "L" shape, on the current via Corte Vecchia and piazza della Cattedrale, and was rebuilt following a fire in the fourteenth century by the architects Ridolfi and Naselli.

Only with Alfonso I, there was the representation of theatrical performances in dedicated permanent spaces, however coexisting for a long time with the temporary staging of non-specific places. Adjacent to the Great Hall, in a building that no longer exists, Ludovico Ariosto created a first theatre in 1531 in order to represent his plays at court, however also destroyed by a fire the year following its opening. In this regard, see the documents reported in the apparatuses. The data known relating to this structure are insufficient to hypothesize an accurate reconstruction. We will therefore limit ourselves exclusively to locating it.

Not many years later, in 1565, under Alfonso II, the Great Hall was rebuilt by the ducal architect Galasso Alghisi. The building stood northwest of the courtyard, overlooking the so-called "caneva", that is the cellars of the building, between the current via Garibaldi and via Corte Vecchia, and was connected to it through a loggia. Sardi reports its dimensions, perhaps excessive, of 82×16.4 m; Aleotti describes it in the plans of Ferrara of 1605 and of 1611, reproduced in the following images. The Great Hall is also recognizable, according to recent studies, in a drawing from Archivio di Stato di Modena, anonymous and without exact dating, also reported, which shows that only half was occupied by the actual theatre, still semi-stably set up. Indeed, different reconfigurations were possible for events and banquets, as suggested by the same partial occupation of the areas. The theatre was divided into stage and spaces reserved for the public as il follows: audience, two side boxes connected to the stage and semi-cyclic steps that anticipate the toothed plans of modern theatres, designed to improve visibility. The Hall was subjected to many changes, including structural ones.

Previously, we mentioned the coexistence of temporary staging of spaces not originally designed to host theatrical performances; in particular, in addition to the courtyard of the Ducal Palace, those of Pendaglia Palace and Schifanoia were also used. These open spaces were characterized by an urban, public dimension, which the private halls did not possess. In addition to the aforementioned Great Hall, the Hall in Cortevecchia, active until 1583 and the Hall in Giovecca, initially used for the game of pallacorda. Starting in 1577, the building of the Great Hall probably hosted knightly exercises, tournament performances, public comedies and art comedies with paid admission. It was the so-called stables theatre, probably measuring 51×8 m, with a front grandstand along the short side. Certainly active until 1594, it brought the events that were once outdoors to the public inside a building. However, even in this case, there is very little information available to attempt an accurate reconstruction of the building as a whole. With the Este devolution of 1598, the Great Hall became public and, subsequently, was transformed into a Theatre of Comedies, thus experiencing the rivalry, competition, with San Lorenzo Theatre, then the Obizzi. Only in 1700, the structure was completely demolished, after a fire which occurred in 1660.

It can therefore be said, in summary, that the Great Hall takes on a character of autonomy within the more complex system of the building. In it all the main genres of representations come together, in particular, the dimension of the public spectacle in an open urban context is fully absorbed. However, the theatre does not yet have specific permanent characteristics, especially in the internal organization, which, on the other hand, remains modifiable according to the needs.

Next to the ducal courtyard and the Great Hall, there is also a third element subject to progressive transformations related to the theatrical representation within the palace complex. This is the so-called court chapel. The first construction of the same is due to the Marquis Leonello between 1441 and 1444 as a single nave structure, with smooth walls, entirely frescoed and dedicated to Santa Maria della Pace, ceiling on pillars and simple elevation. With Ercole I, in 1476, a new chapel was built called Santa Maria di Corte, today Estense Hall, opening onto the new courtyard. The chapel was one of the very few active in Italy in the service of a prince in which polyphonic music was played for daily liturgical functions. The chronicles contain, in particular, the description of the representation of the Passion which took place inside the chapel on April 20, 1481. Ercole, Eleonora and Monsignor Ascanio Sforza attended it from a loggia above the entrance, the stage was erected in front of the altar. Desecrated after the fire in the Hall of the Dauntless Academicians, which took place on January 3, 1660, the chapel was transformed into a theatre by the architects Pasetti and Gnoli. In 1693, it returned to be a place of worship, dedicated to San Maurelio. The function of the Theatre of Comedies had indeed been taken over by the new Scroffa Theatre. Today, immediately to the west of the access staircase, the portal of the suppressed court chapel is still well preserved, which, recently transformed into a cinema, is currently used for cultural events. The facade on the ancient ducal courtyard in which the former chapel is included is finally concluded on the main floor by eight windows decorated by Frisoni, also dating back to the fifteenth century.

This last episode is less significant in the context of the developing process that we want to describe, as it lacks the characteristics of autonomy that were found in the two previous cases. However, it is an example of the dimension of the sacred in the Estense Theatre, a dimension that almost immediately abandoned the religious sphere to celebrate the prince and his centrality.

The theatrical space of the fifteenth century in Ferrara hinges on the prince, identified as a real mythological character. He is surrounded by what we would now call a literary staff and an executive staff.

The Devolution of Ferrara to the Papal State in 1598 can be combined with the concept of devolution of the Estense Theatre, that is, the progressive removal of the prince from the stage, while leaving the centrality of the theatrical activity unchanged.

2.1 Temporarines and Permanency of the Features of the Theatre Space

There have been many changes to the entire complex from the seventeenth century to today, remember in particular the subsequent transformations of the Hall located in correspondence with the court chapel, first occupied by the Mayor's secretary, then by the Danutless Academy, by the Municipal Art Gallery in 1836, by the Council and finally by the Town Council. Between 1923 and 1927, the palace was also the subject, rather than a restoration, of a real reconstruction in the fourteenth-century style based on a project by Borzani, that was very much disputed. However, these subsequent phases did not affect the theatrical sites of the complex, whose definitive decline can be said to have occurred as early as 1700 with the aforementioned demolition of the last vestiges of the Great Hall.

Ferrara is undoubtedly a privileged scenario for study (Fig. 2.1), the events described so far and relating only to the Municipal Palace are indeed only the tip of an iceberg. The performances in the courtyards and squares, set up temporarily, have produced significant transformations in the urban fabric up to the formation of autonomous theatrical organisms, which, in some cases, have also further modified their structure and their function, undergoing their influence and influencing the changes in the city.

Among the main sources describing these variations, there are numerous collections of maps, drawings and papers, kept primarily at Biblioteca Ariostea, among which the most significant are mentioned: *Topografie della città e provincia di*

Fig. 2.1 Giovan Battista Aleotti—Ferrara, 1605, Stampa, 527 × 740 mm, Dedicata a Francesco Borghese [Biblioteca Comunale Ariostea, Serie XVI-63]

Ferrara, Raccolta di stampe e disegni della città e stato di Ferrara con altre città confinanti come pure di diverse stampe e disegni de fiumi e del modo di regolare il Reno, Raccolta di Tavole disegnate o possedute da Giovan Battista Aleotti, Carte corografiche generali e particolari dello Stato di Ferrara e descrittioni compendiose a ciascuna carta.

From the study of these, the architect Carlo Bassi highlighted *the extraordinary geometry* [which] *regulates the urban structure of Ferrara within the walls, welding the Middle Ages and Renaissance into an inseparable unity*. It is no coincidence that Ferrara was defined by Burckhardt[5] as the first modern city in Europe.

Symbol of this unity, as well as the identification of the city with its theatrical scene, is the famous representation contained in the Sardi Code, in which the square between the cathedral and the Ducal Palace is the focal point. Via delle Volte is the matrix path of the city, which developed north of the axis of via Garibaldi—via Corte Vecchia, following the progressive withdrawal of the waters of the river Po and moving the physical center of the town right near the cathedral.

We then refer to the main events that have affected the city spaces for the show with a close link with the complex of the Municipal Palace (Figs. 2.2 and 2.3) and the development of the urban fabric in the areas adjacent to it, highlighting once again the characteristics of temporary and permanence.

There were two main theatres in the eighteenth century, the Bonacossi then Ristori, inaugurated in 1662 and the Scroffa Theatre, which opened its business in 1692. In particular, the latter was born to replace the theatre built in the former court

Fig. 2.2 Piazza municipale, Ferrara [photo by the author]

2.1 Temporarines and Permanency of the Features of the Theatre Space

Fig. 2.3 Town hall, Ferrara—details [photo by the author]

chapel, which returned to its ancient function. It was therefore the new theatre of comedies in Ferrara until 1787, the year in which opera began to be represented and the comedy shows were transferred to Bonacossi. The advent of the Municipal Theatre determined the closure of the Scroffa and its subsequent demolition in 1810. An album of drawings of the Scroffa is preserved in the Archivio Storico Comunale, probable representations of the interventions that were necessary a few years after its first opening, by the architect Francesco Mazzarelli. The main feature of this theatre, as also appears from the Bonacossi geometric survey made by Tosi before 1840, is the wooden structure. Although both, unlike the places dedicated to entertainment in the Municipal Palace, have achieved greater autonomy as buildings and have arisen with the main function of theatre, their internal configuration still has an exquisitely temporary, variable character. The Bonacossi Theatre, with the interventions of the 40s and 80s of the nineteenth century, has instead consolidated its structure. The characteristics of the new theatre, whose decorative apparatus was carried out by Migliari, are furthermore detailed in the *Manifesto for the rebuilding of the Bonacossi Theatre*. The construction works of 1881, entrusted to Carlo Netti, were related to the introduction of gas lighting, the heating system and fire protection devices; the stalls and the foyer were also raised. The nineteenth-century theatre then gave way to the Cinema Theatre Ristori, inaugurated in 1922. Since 2004, the year of the last performances, until today it has been interested in the opening of a new construction site that will transform it into a residential building. The consolidated node of the hall is thus transformed into an open distribution courtyard, inaugurating a new phase, that of a modern building, almost backwards from the common trend of the building specialization process.

The developing process of the modern space for the show starting from some typical characteristics of temporary installations in places that are not specialized at first, and then gradually more and more autonomous, takes place, it was observed, in different contexts. The story of the Ferrara Municipal Palace has been described so far, whose courtyard and internal and neighboring structures are in themselves identifiable and include both the context of the house, that of the building, and that of the urban square.

We therefore want to analyze the case studies of the Cornaro House, the Palazzo Vecchio Theatre and the Capitoline Theatre, which in the process can be assimilated to the Este story but, unlike this, developed in contexts on an architectural and urban scale, whose margins are more strictly defined.

Alvise Cornaro in Padua, thanks to the figure of Giovanni Maria Falconetti, built his house, of which the odeon and the loggia do not represent today parts of a larger and unfinished project, but rather precious remains of what was built at the time. In the in-depth sheet, a reconstructive scheme of the use of the courtyard is proposed, indicating the areas respectively occupied by the scene and by the spectators.

The purely private dimension also belongs to the theatre set up by Giorgio Vasari in the Salone dei Cinquecento in Florence, a real node connecting the two wings of the Palazzo della Signoria, each articulated around an uncovered internal courtyard. From the Memorial of Girolamo Seriacopi, Superintendent of the Castle, and from the accounting documentation it is clear that it was built in 1565 on the occasion of the wedding of Francesco de' Medici and that, for about 24 years, it influenced all subsequent performances. Vasari's theatre consisted of a cavea with six tiers of steps, each about 2 m high; the auditorium was connected to the stage, located at noon at the same height as the first step; the royal box was in the center of the stalls. The theatre represented a real advancement in scenographic as well as in architectural technique. In fact, there was a rational project of the annexed spaces, serving the stage, such as the backstage, the understage, accessible through hatches and the attic, up to the highest point of the hall, placed in communication with the floor above. The scene, about 12 m, was also framed by the scenic arch, consisting of Corinthian columns about 6 m high and the architrave. These intuitions were probably developed thanks to the continuous practice and presence at court of the scenographer Buontalenti and Stefano di Piero Parigi, wood craftsman, from whose family Alfonso and Giulio would descend. These professions were very significant if one thinks of the continuous remodeling and reconfiguration of the hall, which could be combined with a modern studio, set up according to the needs of the scene. The Medici Theatre of 1589, built in the Uffizi Gallery according to a design by Buontalenti, was made up of about sixty wooden steps probably recovered from the Palazzo della Signoria.[6] After such episodes we generally witness in Florence, as well as in Ferrara, a progressive removal of theatre from palace, think of the events of the Cocomero Theatre, inaugurated in 1650, of the Teatro della Pergola starting from 1652, up to the Teatro Nuovo or della Pallacorda, managed by the Accademia

2.1 Temporarines and Permanency of the Features of the Theatre Space

degli Impavidi, and at the Teatro di Borgo Ognissanti of the Accademia dei Solleciti at the end of the eighteenth century.

Finally, a brief mention is made of Pietro Rosselli's Capitoline Theatre, passing from the architectural to the urban scale. This case also summarizes some aspects belonging to the previous ones. It is indeed an open courtyard whose limits are however built from scratch, as well as a temporary structure located in an urban context such as the piazza del Campidoglio. The only existing graphic source is to date the drawing contained in the Coner Code, quoted in Florentine arms, which seems to include thick walls and five orders of steps. The drawing is in all likelihood a project; the realization, described in the chronicles of the time, was different. The theatre was built in 1513 in honor of Giuliano de' Medici, almost entirely in wood, with seven tiers of steps, however reduced in size and covered by a huge veil of cloth. It joined the Palazzo dei Conservatori on one side; the opposite side, corresponding to the scene, was instead near Palazzo Senatorio; it occupied a good part of the square; with an estimated plan size of about 35 × 31 m it could accommodate between 800 and 1000 spectators in the stalls, between 700 and 800 in the benches of the cavea; the height and the architectural order are still uncertain, only the plan having been received. There were five access arches, of which the largest is the central; at the back of the scene five more doors were reproduced. The front was similar to the Roman triumphal arches, representative of the heroic world they wanted to evoke, with scenes painted between the arches.

Ultimately, we are witnessing a process of identifying and characterizing the space for the show in which the autonomy of the built remains partial and the main distinctive features are temporary.

2.2 The Representation of the Renaissance and Baroque Festival References

The Renaissance and Baroque fest was a unicum, a representation whose characteristic par excellence was the intermediality, the temporary coexistence of all the arts and all the means to make it real, as it would happen later in the opera.

In the following, we want to deepen, in the derivation from the sacred and the profane, not so dissimilar to each other especially in terms of spatial organization, some of the most recurrent temporary characters in representations.

While indeed we have seen, on the one hand, the process of autonomy of theatre, which therefore acquires an independent space exclusively dedicated to the function, progressively moving away from the court, on the other, it has been highlighted as the structure of the distinctive characteristics of the area for the performance is still too closely linked to the occasion. However, often these temporary organizations have proved to be absolutely, and also disconnected from the specific context in which they were created, forerunners of some of the main characters of the modern space for the show.

What it seems certain is that, at least until the 1980s, neither in this room nor in other rooms of the ducal residence a real permanent theatre was built. The extreme mobility of the performance venues inside and outside the building, and the provisional nature of the related structures remains one of the most evident peculiarities of the Ferrara theatre of the time. The structures, especially in the part intended to accommodate spectators (boxes and stairways) consisting of a sort of modular elements to be assembled in the required dimensions, were continually modified, reused and adapted in different areas, even not specifically theatrical.[7]

Sergio Monaldini's statement fully summarizes the Ferrarese context of the fifteenth and sixteenth centuries. Indeed, often elements such as boxes and steps were temporarily sold and reused by the noble families of the city, who represented shows privately, especially after the devolution, in a context of widespread patronage. A primary role was therefore played by the carpenters, among whom, it has been mentioned, also the greatest scenographers of the time, as well as skilled conductors of the theatres themselves, emerged.

The reconstructions of the organizational structures are therefore possible mainly thanks to historical chronicles and in some rarer cases, as for the Capitoline Theatre, thanks to graphic representations. It was also observed that for the Great Hall of the Ducal Palace the arrangement of the boxes is an anticipation of the modern theatre.

The serial vocation of the wooden elements is emphasized, progressively replaced by plastic-wall components, accentuating the increasingly marked organicity of the building.

The performances took place in the Renaissance and Baroque periods, throughout our peninsula, mostly during the Carnival period, between January and February, with unfavorable climatic conditions. Therefore, an attempt was made to close the courtyards with temporary wooden structures or with cloths and tapestries, as well as to heat them. Concrete needs, therefore, fueled the gradual transition to covered spaces, specialized and intended only for the specific function. Covering the place for the show does not only mean identifying it by delimiting it but also ennobling it, making it representative by developing the identity characters of the communities, according to two main languages: the religious ceremonial and the chivalrous ceremonial, first linked to different spaces, which then tend to assume in time characteristics that are increasingly similar to each other, see the example of the Great Hall mentioned above.

We read in Vasari's Life of Cecca: "*La piazza di San Giovanni si copriva tutta di tele azzurre, piene di gigli grandi fatti di tela gialla e cucitivi sopra; e nel mezzo erano in alcuni tondi, pur di tela e grandi braccia dieci, l'arme del popolo e Comune di Firenze, quella de' capitani di Parte Guelfa; ed intorno negli estremi del detto cielo, che tutta la piazza, comecchè grandissima sia, ricopriva, pendevano drappelloni pur di tela, dipinti di varie imprese, d'armi di magistrati e d'Arti, e di molti leoni, che sono una delle insegne della città. Questo cielo, ovvero coperta così fatta, era alto da terra circa venti braccia; posava sopra gagliardissimi canapi attaccati a molti ferri, che ancor si veggiono intorno al tempio di San Giovanni, nella facciata di Santa Maria del Fiore, e nelle case che sono per tutto intorno alla detta piazza, e fra l'un campo e l'altro erano funi, che similmente sostenevano quel cielo; che per tutto era*

2.2 The Representation of the Renaissance and Baroque Festival …

in modo armato, e particolarmente in su gli estremi, di canapi, di funi, e di soppanni e fortezze di tele doppie e canevacci, che non è possibile immaginarsi meglio. E, che è più, era in modo e con tanta diligenza accomodata ogni cosa, che, ancorachè molto fussero dal vento, che in quel luogo può assai d'ogni tempo, come sa ognuno, gonfiate e mosse le vele, non però potevano essere sollevate né sconce in modo nessuno. Erano queste tende di cinque pezzi, perché meglio potessero maneggiare; ma, poste su, tutte si univano insieme e legavano e cucivano di maniera, che pareva un pezzo solo. Tre pezzi coprivano la piazza e lo spazio che è fra San Giovanni e Santa Maria del Fiore; e quello del mezzo aveva, addirittura delle porte principali, detti nodi con l'arme del Comune; e gli altri due pezzi coprivano le bande, uno di verso la Misericordia, e l'altro di verso la Canonica ed Opera di San Giovanni".

Another important aspect is the substantial coexistence of the city as a theatrical place that occasionally modifies its squares and buildings making them places linked to the ephemeral, through the use of temporary devices, with permanent theatres, specialized closed buildings. Representations are in fact still today in various examples linked to the urban context. The Spoleto Festival dei Due Mondi is mentioned in Fig. 2.4, now in its 64th edition. The Festival enhances live representation, performance in every artistic sector, both in traditional places such as theatres, halls, auditoriums, cinemas and in open spaces such as squares and gardens. In particular, the culmination of the event is still today constituted by the shows in piazza Duomo. The square stands at the edge of the city and at the same time, despite the physical margins given by the conformation of the land, effectively inserted into the landscape. The difference in height describes the perimeter, creating an intimate space, although with an extension of over 3000 m^2. The nodal axis runs from via dell'Arringo to the Duomo, which, with its arcades, forms the natural scene. Additional architectural highlights on the square are the Caio Melisso Theatre, the Church of the Golden Manna and the Teatrino delle 6, structures that make up the former Palazzo della Signoria, as well as Palazzo Bufalini and Palazzo Racani Arroni. The disposition of the public in sectors and that of the stage, not very high compared to the floor level, has consolidated over time, reaffirming the centrality of the city as a theatrical place, stimulus always on and alive.

Another type that partially summarizes what has been described up to now is that of the outdoor stable theatre. Also in Ferrara in 1627 the Francesco Guitti Theatre was inaugurated in the courtyard of San Pietro Martire in Pilotta. It provided about 3500 seats, an elevated distribution corridor, flanked by four staircases, a real distributional prefiguration of the modern theatre, as well as the anticipation of the bell shape to increase visibility and, finally, two side doors between the bell and the stage.

In conclusion, the Dauntless Academy Theatre, then Obizzi, then Tosi-Borghi and then Verdi is worthy of further study. The first floor plan of Danutless Academy Theatre, dating back to the early seventeenth century and shown below, is by Giovan Battista Aleotti. It is said that Aleotti *"reduced the granary near the church of San Lorenzo into a beautiful arena"*. This theatre is a real anticipation of the Farnese Theatre in Parma, but there are also similarities with Serlio, Bertani, Palladio and Scamozzi. The first ever opera performance was represented in it, Claudio Monteverdi's *Orfeo*, it had dimensions of approximately 21.3 × 36.5 m, it was

Fig. 2.4 Piazza del Duomo in Spoleto during the Festival dei Due Mondi (2018) [photo by the author]

2.2 The Representation of the Renaissance and Baroque Festival …

probably equipped with a proscenium. The theatre was then purchased by the Obizzi in 1640, thus losing the link with the Academy. The architect Carlo Pasetti then transformed it into "modern architecture", merging the theatrical venue for chivalrous performances with the hall theatre. A large free space was created in the center, surrounded by the stepped cavea, tribunes with several superimposed bridges arranged circularly along the perimeter of the stalls, the stage with an architraved facade, all favoring a frontal view of the scene. These characteristics can be said to be precursor in all respects since the Aleotian order, then reiterated in Parma.

Prisciani in *Spectacula*, the first treatise in vernacular dedicated to theatre, a Vitruvian and Albertian synthesis, as well as a symbol of the rediscovery of classical theatre in the Ferrara context, makes the word *theatro* descend from visorio, a place where it is possible to see and be seen, according to the interpretation of Cassiodorus. Prisciani also distinguishes the five necessary parts of theatre: *"lo expedito sino de la discohoperta area mediana, le gradazione atorno atorno la area, lo exagerato pulpito, el portico nel supremo et interiore ambito, et el cohoperto"*. These elements are performed in the Aleottian theatre.

The theatre was destroyed by a fire in 1679, its remains were not demolished or removed for about 130 years, after which it was decided to free the area to use it for the horse market, taking into account that the southeastern side of the square overlooked the church of San Lorenzo, no longer existing today.

On those same places, an open summer arena was built again, inaugurated in June 1857, based on a project by the architect Foschini. Just 6 years later, in 1863, the theatre was covered, thus determining the transition from daytime theatre to theatre for performances in general, also possible in winter. However, it maintained a vocation for daily representation while still retaining a glass roof. In a letter from Angelo Borsari to the mayor of 1863, we read: *"the covering of the stalls consists of various iron cavities secured to the internal fulchers of the scaffolding, on which fir boards rest on the outside covered by painted canvas, they include an area around the curve, and in the middle an opening covered with common sheets of crystal supported by a wooden frame, and defended by an iron mesh"*.[8] This roof was subject to a collapse in 1871 and restored according to a design by Engineer Calzoni. A letter from the Chief Engineer Borsari to the mayor shows that this project included: *"an iron and timber roof on cast iron columns to cover the stalls which can be supported independently of the rest of the theatre; and its columns should indeed support the parapets of the loggias and the attics adhering to them"*.[9] Ultimately, we are witnessing a real knotting of a space originally open for the show. On May 7, 1913, the Verdi Theatre was inaugurated; abandoned in recent years, today it is subject to a restoration that will make its spaces accessible to the city again, but with new and different functions.

2.3 Summary

We can say that we have analyzed, through the case studies of the Municipal Palace of Ferrara, of Cornaro House, of the theatre of Palazzo Vecchio and of the Capitoline Theatre, the process of characterization of the modern space for the show.

The autonomy of the theatre building is only partially achieved, theatres do not yet have specific permanent characteristics, especially in the internal organization, which, on the other hand, remains modifiable according to the needs.

It was then underlined how temporary forms of space organization were precursors of some of the main characters of modern theatres.

Finally, through the case study of piazza Duomo in Spoleto as part of the Festival dei Due Mondi, the centrality of the city as a theatrical place, an ever-burning and lively stimulus for transformation, outside an exquisitely chronological order and instead to the internal of a constant process.

2.3 Summary

TEATRO ALL'APERTO CODE2.01

Town Hall, Ferrara - 2017 - reconstructive hypothesis of the open-air theatre

TYPOLOGICAL SCHEME

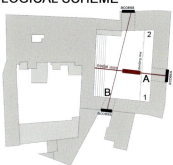

LEGEND

A SCENE
1 balcony of Ercole
2 Paradise
B AUDIENCE

SUMMARY OF THE MAIN SELECTION CHARACTERS

place_**FERRARA**
level of the event_**regional**
site of architectural interest_**Yes**
UNESCO site_**Yes**
year_**1487**
building_**temporary theatre**
construction_**wooden**
features_**scene distinct from the area dedicated to spectators**
keyword_**organism**

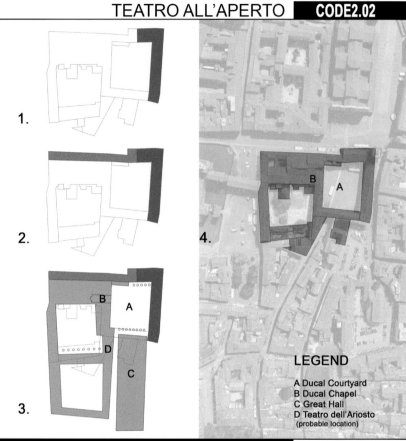

SUMMARY OF THE MAIN HISTORICAL PHASES

1_The original nucleus of the Ducal Palace, 13th century

2_The Ducal Palace rebuilt after a fire, 14th century

3_The Ducal Palace after the completion interventions wanted by Ercole I, 15th-16th century

4_The Town Hall today

2.3 Summary

CORTILE DI CASA CORNARO — CODE2.03

The loggia and the odeon - 2017

TYPOLOGICAL SCHEME

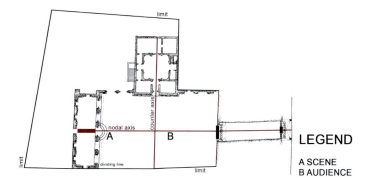

LEGEND
A SCENE
B AUDIENCE

SUMMARY OF THE MAIN SELECTION CHARACTERS

place_**PADOVA**
level of the event_**local**
site of architectural interest_**Yes**
UNESCO site_**Yes**
year_**1524**
building_**temporary theatre**
construction_**wooden**
features_**scene distinct from the area dedicated to spectators**
keyword_**enclosure**

TEATRO DI PALAZZO VECCHIO — CODE2.04

Salone dei Cinquecento 2017 (Teatro di Palazzo Vecchio - 1565)

TYPOLOGICAL SCHEME

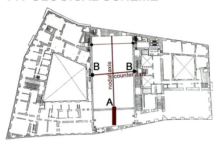

LEGEND

A SCENE
B AUDIENCE

SUMMARY OF THE MAIN SELECTIONS CHARACTERS

palce_**FIRENZE**
level of the event_**national**
site of architectural interest_**Yes**
UNESCO site_**Yes**
year_**1565**
building_**temporary theatre**
construction_**wooden and brick**
features_**scene distinct from the area dedicated to spectators**
keyword_**organism**

2.3 Summary   45

TEATRO CAPITOLINO | CODE2.05

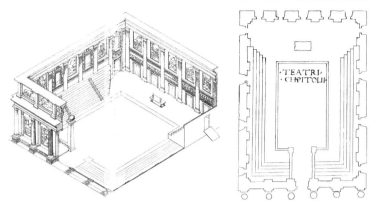

Teatro Capitolino: reconstructive hypotesis (Bruschi, 1969), plan from Codice "Coner", f. 16r - 1513

TYPOLOGICAL SCHEME

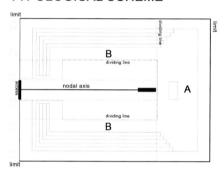

LEGEND

A SCENE
B AUDIENCE

SUMMARY OF THE MAIN SELECTION FEATURES

place_**ROMA**
level of the event_**national**
site of architectural interest_**demolished**
UNESCO site_**Yes**
year_**1513**
building_**temporary theatre**
construction_**wooden**
features_**scene distinct from the area dedicated to spectators**
keyword_**enclosure**

PIAZZA DEL DUOMO A SPOLETO — CODE2.06

Piazza del Duomo di Spoleto - Festival dei Due Mondi - 2016

TYPOLOGICAL SCHEME

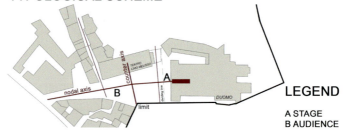

LEGEND
A STAGE
B AUDIENCE

SUMMARY OF THE MAIN SELECTION CHARACTERS

palce_**SPOLETO (PG)**
level of the event_**international**
site of di architectural interest_**Yes**
UNESCO site_**Yes**
year_since **1958**
building_**temporary theatre**
construction_**wooden and still**
features_**scene distinct from the area dedicataed to spectators**
keyword_**organism**

2.3 Summary

COMPARATIVE TABLE CODE 2.27

SINGLE SCENE DISTINCT FROM THE AREA DEDICATED TO SPECTATORS
ENCLOSURE
OPEN COURTYARD, INCOMPLETE MARGIN LINE, DOUBLE AXIALITY

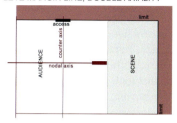

SYNCHRONIC

ENCLOSURE
OPEN COURTYARD, COMPLETE MARGIN LINE, DOUBLE AXIALITY, SEATS OPPOSED TO THE STAGE
HALL INSIDE THE PALACE, COMPLETE MARGIN LINE, DOUBLE AXIALITY, SEATS OPPOSED TO THE STAGE

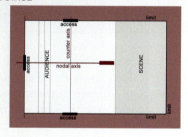 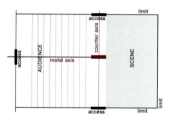

ENCLOSURE
OPEN OR PARTIALLY COVERED TEMPORARY BUILDING, COMPLETE MARGIN LINE, SINGLE AXIALITY, SEATS CONNECTED TO THE STAGE

Notes

1. «*Nel teatro medievale, si è detto, le costruzioni sceniche, per quanto grandiose e articolate, sono superfetazioni delle architetture e delle aree urbane, sono luoghi provvisori che costruiscono il proprio senso sopra e negli interstizi degli spazi codificati della città. E infatti, coerentemente, il problema dell'edificio teatrale si pone ad una cultura svincolata da questi presupposti ideologici,*

come quella rinascimentale, quando anche il teatro stesso viene riconosciuto, perseguito e istituito senza ambiguità come pratica culturale non solo legittima ma dignificante e necessaria. [...] Il recupero di un modello antico, falso storicamente ma credibile e in buona fede a partire dai dati a disposizione, si può configurare allora in realtà più come una proiezione in avanti, una mossa da giocare al presente, in cui trova risposta un bisogno di teatro non più differibile, che un tentativo di ricostruzione filologica». Allegri L, *Teatro e spettacolo nel medioevo*, Editori Laterza, Roma 1988.

2. Zamboni A (1487) Diario Ferrarese.
3. The first performance that took place in the courtyard of the Ducal Palace was the Plautian opera *Menaechmi*, translated into vernacular, with the scenes by Pellegrino Prisciani. *Amphitryon*, object of the description, was instead staged on February 5, 1487.
4. The staircase was then rebuilt in 1880.
5. Jacob Burckhardt 1860: 55.
6. The Medici Theatre fell into disuse after the death of Parigi, it was temporary occupied by the Senate and today it is used as an exhibition gallery.
7. *«Quel che sembra certo è che, almeno sino agli anni'80, né in questa sala né in altri ambienti della residenza ducale sia stato edificato un vero e proprio teatro stabile. L'estrema mobilità dei luoghi di recita fuori e dentro il palazzo, e la provvisorietà delle relative strutture rimangono anzi una delle peculiarità più evidenti del teatro ferrarese dell'epoca. Le strutture, specie nella parte destinata ad accogliere gli spettatori (palchi e scaloni) costituite da una sorta di elementi modulari da comporre nelle dimensioni richieste, venivano continuamente modificate, riutilizzate ed adattate in ambiti diversi, anche non specificatamente teatrali».* Monaldini A (2002) I teatri della commedia dell'arte. Le prime sale, il teatro della Sala Grande, l'ex Cappella Ducale. In Fabbri P, I teatri di Ferrara, Lim Editrice, Lucca: 5–6.
8. FEasc TS b. 56.
9. Ibid.

References

Bentini J, Spezzaferro L (1987) L'impresa di Alfonso II. Saggi e documenti sulla produzione artistica a Ferrara nel secondo Cinquecento. Nuova Alfa Editoriale, Bologna

Andrews R (1993) Scripts and scenarios. The performance of comedy in Renaissance Italy. Cambridge University Press, Cambridge

Bertieri MC (2012) I teatri di Ferrara. Il Tosi—Borghi (1857–1912). LIM Editrice, Lucca

Bortolotti L (1995) Le stagioni del teatro. Le sedi storiche dello spettacolo in Emilia-Romagna. IBC, Bologna

Bosi G, Mazzanti MB, Mercuri AM, Torri P, Grandi GT, Accorsi CA, Scafuri F (2006) Il Giardino delle Duchesse del Palazzo Ducale Estense di Ferrara da Ercole I (XV sec.) ad oggi: basi archeobotaniche e storico-archeologiche per la ricostruzione del giardino. In: The archaelogy of cropfields and gardens pp 103–128

Cavicchi C, Ceccarelli F, Torlontano R (2003) Giovan Battista Aleotti e l'architettura. Casa Editrice Diabasis, Reggio Emilia

Cruciani F (1969) Il teatro del Campidoglio e le feste romane del 1513. Il profilo, Milano

Dalla Negra R, Ippoliti A (2014) La città di Ferrara, architettura e restauro. Città di Castello

Fabbri P (2002) I teatri di Ferrara. LIM Editrice, Lucca

Fiocco G (1958) La casa di Alvise Cornaro. Storia e Letteratura (72):69–77

Gelichi S (1992) Ferrara prima e dopo il Castello. Testimonianze archeologiche per la storia della città. Spazio Libri Editori, Ferrara

Maione C (2010) La devoluzione del teatro estense. Schifanoia (38–39):257–260

Maione C (2011) Lo spazio del sacro nel teatro di Ercole I d'Este. Schifanoia (40–41):147–156

Maretti A (1992) Dal teatro del principe alla scena dei virtuosi. Indicazioni sul mecenatismo di Mattias De' Medici (1629–1666). Medioevo e Rinascimento VI(III):195–210

Peron M, Savioli G (1986) Ferrara Disegnata, riflessioni per una mostra. Ferrara

Strappa G (2014) L'architettura come processo. Il mondo plastico murario in divenire. Franco Angeli, Milano

Testaverde AM (1992) Informazioni sul teatro vasariano del 1565 dai registri contabili. Medioevo e Rinascimento VI(III):83–96

Ventrone P (2016) Teatro civile e sacra rappresentazione a Firenze nel Rinascimento. Le Lettere, Firenze

Documents

Matteo Florimi, Ferrara, 1598, Incisione, 426x544mm, In alto a sinistra lo stemma Aldobrandini. Rivista da Giuseppe Capocaccia [Biblioteca Comunale Ariostea, Serie XIV-4]

Giovan Battista Aleotti, Pianta come andrebbe fatta la Fortezza se ritornasse il Po navigabile, 1605, Disegno a penna, 511x709mm, Copia a penna dell'originale del 1597. Sul retro nota manoscritta a matita di Eugenio Righini, 29 aprile 1932 [Biblioteca Comunale Ariostea, Serie XIV-5]

Mattio Cadorin, Pianta di Ferrara, Padova, 1669, Stampa, 130x188mm [Biblioteca Comunale Ariostea, Serie XIV-7A]

Gerolamo di Novo dis., Giovanni Orlandi inc., Ferrara, Roma, 1602, Incisione, 573x531mm, Rivista dal Signor Giuseppe Capocaccia. A penna è aggiunta la legenda delle cose notevoli [Biblioteca Comunale Ariostea, Serie XV-60]

Giovan Battista Aleotti, Ferrara, 1605, Stampa, 527x740mm, Dedicata a Francesco Borghese [Biblioteca Comunale Ariostea, Serie XVI-63]

Giovan Battista Aleotti, Ferrara, Disegno a penna, 1255x936mm, Disegno autografato dall'autore [Biblioteca Comunale Ariostea, Serie XVI-64]

Ruggiero Moroni, Ic[o]hnographia della Piazza di Ferrara da Ruggiero Moroni fatta di commiss.ne del Ill.mo et Ecc.mo Sig.r Marc.se Cesare Calcagnino Giudice de Savi e suo maestrato per il nuov ordine e riforma delle botteghe per la piazza e Livellari […], 1618, Disegno a penna e a colori, 1408x1460mm [Biblioteca Comunale Ariostea, Serie XVI-73]

Teatro Bonacossi. Proprietà palchi comunali [Archivio Storico Comunale di Ferrara, Carteggio Amministrativo, secolo XIX, Teatri e Spettacoli, Varie, busta n. 70]

Teatro Scroffa, Montecatini, Filodrammatico [Archivio Storico Comunale di Ferrara, Carteggio Amministrativo, secolo XIX, Teatri e Spettacoli, Varie, busta n. 71]

Disegno Teatro Obizzi (interno e esterno). (Stampe). [Biblioteca Comunale Ariostea, Fondo Antolini, busta n. 83]

1850 Lavori al Palazzo Comunale. 1898 Restauro al tetto del Palazzo Comunale. Concorso dei comproprietari. 1888 Restauro ai prospetti del Palazzo Comunale. 1802 Lavori eseguiti alla Locanda del Cavalletto. [Archivio Storico Comunale di Ferrara, Carteggio Amministrativo, secolo XIX, Fondi Comunali, Palazzo Civico, busta n. 67]

Palazzo Comunale di Ferrara [Archivio Storico Comunale di Ferrara, Mappe, stampe e disegni, 3/A dal 232 al 332]

Palazzo Comunale di Ferrara [Archivio Storico Comunale di Ferrara, Mappe, stampe e disegni, 3/A bis dal 333 al 404]

Palazzo Comunale di Ferrara [Archivio Storico Comunale di Ferrara, Mappe, stampe e disegni, 4A dal 406 al 532]

1905, Fotografo: L. CAVALLINI, Note Fronte: Eseguita dal Sig. Luigi Cavallini per incarico del Locale ufficio del Genio Civile Aprile 1905, (G. Agnelli?) [Biblioteca Comunale Ariostea, Fotografie, O.89.4 f.n. 225]

1905, Fotografo: L. CAVALLINI, Note Fronte: Eseguita dal Sig. Luigi Cavallini per incarico del Locale ufficio del Genio Civile Aprile 1905, (G. Agnelli?), *Timbro B.C.A. [Biblioteca Comunale Ariostea, Fotografie, O.89.4 f.n. 226]

1905 –1908, Fotografo: L. CAVALLINI, Note Fronte: Eseguita dal Sig. Luigi Cavallini per incarico del Locale ufficio del Genio Civile Aprile 1905, (G. Agnelli?), *Timbro B.C.A. [Biblioteca Comunale Ariostea, Fotografie, O.89.4 f.n. 224]

1910–1912, Note Fronte: Fotografia di un quadretto a colori nel quale trovasi la seguente iscrizione. Il Castello di Ferrara nel 1400 tolto da un originale da trovarsi nel Palazzo del cardinale Luigi d'Este a Tivoli—Roma, *Foto su supporto rigido cartoncino, *Timbro in basso a sinistra dell'ufficio u.p.r. del Comune di Ferrara [Biblioteca Comunale Ariostea, Fotografie, O.89.4 f.n. 132]

1915–1916, Ferrara: alzato della parte est di città con il sobborgo di S. Giorgio (Episodio Guerresco), ALLEGATO: Cart. Di supp., NOTA: Recto cartoncino [Biblioteca Comunale Ariostea, Fotografie, O.89.4 f.n. 184]

1916–1918, Fotografo: S. Della Valle, Ferrara: da Silog. nel cod. Sardi: annotazioni istor. della Bibl. Estense, 5: Modena, Segn. a.F.13.?, Allegato: cartoncino di supporto, note: verso cartoncino, recto fotografia [Biblioteca Comunale Ariostea, Fotografie, O.89.4 f.n. 59]

Chapter 3
Theatre as Solidification

Rediscoveries, Interpretations, Inventiones

Abstract An undoubted role in the affirmation of the space for the modern spectacle was played by the studies of classical texts. In the following, we want to investigate the aspects of Vitruvianism, rediscovery, interpretation and invention, persisting in not only Palladian but also Scamozzi's work. In fact, we can rightly speak of solidification, which occurred at the end of the sixteenth century with two main projects that have come down to us and that have in fact marked the developing process of theatre: the Olimpyc in Vicenza and the Teatro all'Antica di Sabbioneta. The two figures of Palladio and Scamozzi represented the critical conscience, the original contribution to the developing of the type, the invention, which arrives at the end of the very long phase that has been described so far, of temporariness, research and identification of the autonomy of the theatrical space in close relationship with the urban fabric; phase that has been seen to last until actuality and coexists with the theatre building. The updating of the type obtained thanks to these authors and architectural works also opens, precisely with solidification, a new condition, that of independent, permanent, repeatable theatre, openly confronting the latest example in this sense constituted by classical theatre, and marking its overcoming. If we want, modern theatre is now ripe to deal with an interrupted tradition. In conclusion, there are two of the most representative results of the renovation, the Court Theatre of Mantua and the Farnese Theatre of Parma.

Keywords Solidification · Andrea Palladio · Vincenzo Scamozzi · Vicenza · Sabbioneta

3.1 Reinventing Classical Theatre

As it is known, Palladio, before carrying out the project for the Olympic Theatre (Fig. 3.1), studied for a long time the design of Raphael's open-air theatre for Villa Madama, which was never built, he completed surveys of classical buildings such as the Berga Theatre and the amphitheatre of Verona, he also took care of temporary installations in Vicenza (1561–1562), commissioned by the same Olympic Academy founded in 1555 and in Venice (1565). The Olympic is therefore the third theatre built by Palladio, the first permanent and still existing; it was designed in the year of

© Springer Nature Switzerland AG 2022
S. Clemente, *Reconstructing Theatre Architecture*, The Urban Book Series,
https://doi.org/10.1007/978-3-030-89968-4_3

Fig. 3.1 Andrea Palladio—Olympic Theatre in Vicenza

3.1 Reinventing Classical Theatre

his death, 1580, in the so-called *loco delle pregion vecchie*, requested in concession by the Academicians to the Municipality of Vicenza, the day after the master's development. A dual and simultaneous realization hypothesis of the scaenae frons of the theatre is the only existing autograph evidence of the work.

As mentioned at the beginning, before the Olympic all the theatres had characters of provisionality, with the exception of the Mantuan court theatre built by Bertani between 1549 and 1551. The first elementary constructive action that qualifies the permanence of the theatre building is the cover, the conformation of which is the subject of debate. The current ceiling painted as a blue sky with clear clouds and the decorated coffered ceiling delimiting the proscenium date back to a refurbishment in 1914, preceded during the nineteenth century by an illusionistic velarium. Although there is no certain evidence regarding the architect's original idea, it is reasonable to assume, Licisco Magnato argues,[1] that the structural conformation of the versurae, equipped with corner columns, was designed to support a so-called "ducal-style" ceiling, consisting of wooden lacunars hanging from the roof trusses, decorated with stucco and tempera, to cover the stage, correctly interpreting the form of the classical theatre, constituting an embryonic proscenium and thus masking the attic towards the Scamozzi scenes.

The proscenium would be configured even more as a key element, coordinating the architectural theme, shielding and not framing, leaving the perspective scenes in the background, devoid of a single focal point. Contrary to what happened in classical theatre, it can nevertheless be affirmed that the Palladian versurae involve in themselves, beyond any reconstructive hypothesis, the union between the stage and the audience, anticipating the nodal character of the modern theatre towards the unification of the architectural organism.

Therefore, on the one hand, we can observe the *Palladio's will to adapt ancient structures to a closed environment, created to meet modern functional needs, respecting the geometric-spatial synthesis of the cavea, the orchestra, the proscenium and the scaenae frons of the Roman theatre*[2]; on the other hand, it can be said that *the novelty arises from re-proposing an organically revived language, to express modern concepts and needs of architecture, contaminating apparently contradictory experiences between them.*[3]

We witness the observance of the Vitruvian treatise, in perfect harmony with the interpretation given by Barbaro, especially in the relationships between the components of the theatrical organism, with the necessary precautions dictated by the need for contextualization in the urban fabric. The theatre has a construction based on Vitruvian proportions, but these refer to an elliptical plan, making it wider than it is long, making the most of the available space, increasing visibility and generating an incredibly wide proscenium, 25×6 m.

On the death of Palladio, the direction of the works was first entrusted to his son Silla (1581), then to Maganza and Scamozzi, in collaboration with Angelo Ingegneri, who played a decisive role in the choice of a tragedy, *The tyrant Oedipus*, for the inauguration of the theatre, with what ensued at the stage level, of adapting the initial idea. However, although Scamozzi's activity was autonomous and significant, the Olympic project can still be considered respected in its essential constitutive lines.

Despite the importance of Palladio's contribution to the theatre type with this project, many of the contradictions highlighted that admirably held together different intuitions, if on the one hand they made the building unique, on the other hand, they decreed its non-repeatability, unless to provide a new and original solution to them. The scaenae frons of correct philological reconstruction coexists with the typical perspectives of the theatrical conception of the school of Peruzzi, Serlio and Vasari and perhaps also with the proscenium, whose invention is ascribable, as described, to the Vasari Theatre at the Medici court. The scaenae frons is precisely the element most linked to classicism, no longer reproduced in subsequent buildings except in the Scientific Theatre by Bibiena centuries later and with the necessary adaptations. Yet it too has a modern character. Palladio interprets Vitruvius' description of the squares of the Romans and the Greeks as very similar to the scenic apparatus of the theatre, the Engineers in fact approach the scaenae frons to the courtyard of a palace, inspired by Cornaro House, especially suited to the tragic representation, with the perspectives depicting pieces of the city; the scaenae frons is therefore an illustrious building, while the proscenium represents its internal courtyard, the square. Still Magagnato about the proscenium: *"in this way it assumes the very role of symbolic synthesis of what the city realistically staged from the perspectives of the Peruzziano-Serliano theatre had been. It is a new way of viewing the constant scenery of the sixteenth-century theatre, which is the idea of the city; in Palladio it is concentrated and absorbed in the looming columnatio, which conforms to all the architectural elements, even secondary aspects"*.[4]

3.2 Vitruvius as a Source of Modern Constructive Pragmatism

Both Palladio and Scamozzi dealt with the translations of Vitruvius, think of that by Fra Giocondo in 1511, by Cesariano in 1521 as well as of the edition by Barbaro of which Palladio himself took care of the graphic representations. Such knowledge, as we have already seen with Palladio and will see in an even more accentuated way with Scamozzi, underline the conscious deviation of the authors from the Vitruvian model, their critical conscience, their grafted paths, contextualized in the urban fabric.

Scamozzi studied the treatise of Vitruvius in the translated and commented version by Daniele Barbaro for the first time in 1574, on the original owned by the Count of Cicognara, enriching it with interesting annotations, as can be seen from the *Catalogo ragionato dei libri d'arte e d'antichità posseduti dal Conte di Cicognara*. In the posthumously published treatise, *Idea of Universal Architecture*, Scamozzi shares the Vitruvian thought of architecture as domina artium and exalts the close link between urban planning and architecture, especially his interventions in the cities of new foundation of Sabbioneta and Palmanova. However, he does not fail

3.2 Vitruvius as a Source of Modern Constructive Pragmatism

to define Vitruvius as an author considered by many to be obscure, difficult and torn. He has a conception of architectural drawing as a strictly mathematical and geometric phenomenon, very close to the current one, he considers the knowledge of monuments and construction techniques very important, recognizing architecture as a universal utility, as well as movement and context insertion.

The institution of the Studio Reformers by the Venetian Magistracy in 1528, the birth of academies and communities of scholars, the Paduan anti-Aristotelianism, the proximity to Galileo and even the humble origins of Scamozzi can all be considered determining factors in the desire for autonomy, innovation and also the subversion of traditions. Establishing an ambivalent relationship with Vitruvius, one wonders, on the threshold of the seventeenth century, what is legitimate to keep updated of the Vitruvian theories, what is still valid of the previous system in the arts and sciences during the Galilean revolution.

In summary, Scamozzi has a practical vision of the world, he uses and refers to the techniques of the ancients only when these are useful, making a selection, carrying into modernity the same pragmatism that contemporaries had recognized in Vitruvius.

The Teatro all'Antica di Sabbioneta (Fig. 3.2) was commissioned to Scamozzi by Vincenzo Gonzaga, Duke of Mantua from 1587 to 1612, also a member of the Olympic Academy where the architect had become supervisor of the Olympic Theatre. The city of Sabbioneta was built starting from 1554 with the Ducal Palace and completed in 1590 with the theatre, in a conformation almost identical to the present one. Vespasian's desire was to give life to a new Rome, and there are numerous

Fig. 3.2 Vincenzo Scamozzi—Teatro all'Antica in Sabbioneta—exterior [photo by the author]

allusions in iconography and architecture, think of the Circus Flaminio and the Circus Maximus reproduced as frescoes in the Sala delle Olimpiadi of the Palazzo Giardino, as well as the presence of elements, such as the column surmounted by the statue of Minerva located in piazza D'Armi, moved during the sack of Rome. A Rome whose model was increased in importance by the spread of prints that replicated it in its ancient splendor and made it much more influential in Europe than it was due to the ruins concretely present in the city fabric; a model therefore rediscovered, interpreted and the subject of new inventions. The motto reported on the fronts of the theatre: *"Roma quanta fuit ipsa ruina docet"* is of clear Serlian ancestry.

It is no coincidence that this is erected on the decumanus of the city, at the confluence of two main urban routes, as it is typical of specialized buildings, halfway between the public and private dimensions, consisting, respectively, of piazza Ducale, with all the representative bodies that overlook it, and Palazzo Giardino.

The theatre is the first example of a modern building not linked to a pre-existing construction, such as the Olympic, and with a majestic autonomous architectural facade, free on three sides, each containing an access.[5]

Further tributes to Serlian treatises are the rectangular plan of the room and the scene, in wood, with a single perspective focus. It has an extension of 36.68 × 11.32 m, the transverse measurement is also repeated in height in order to give the section a square shape. Alongside the recognized academicism of Serlian ancestry, the stage, rather than a physical object, is the result of a perspective elaboration as for Barbaro. Scamozzi emphasizes the universal effectiveness of regular forms, he wants to establish a relationship between the natural universal order perceived by mind and the particular qualities of buildings; the building responds primarily to use, to the characteristics of acoustics and visibility typical of the spectacle of the time, very distant from those of ancient Rome and ancient Greece.

Sabbioneta Theatre is eloquent in its lack of references to Vitruvian principles, it presents itself as a daring invention, as a unique organism of its kind; see, for example, as structuring characteristics at the basis of the continuity of the developing process:

the sloping floor near the auditorium of Florentine origin (now replaced by a flat one);

the subdivision of the places for women, in the loggia (Fig. 3.3), and for men, on the steps; the box of Vespasiano Gonzaga located in the loggia and not near the orchestra or the cavea, as was the case for the Roman senators;

the versurae, the side entrances typical of classicism, reduced to painted walls, only one of which is really equipped with a threshold; the same walls depict Castel Sant'Angelo and the Capitol, replicating a city scene inside;

the orchestra located at the back of the theatre in a completely unconventional position, on the second floor; it should be noted that the reverberation time with the room occupied is identical to that of modern rooms;

the ceiling as a reproduction of the sky, such as to make the building perceived as the theatre of the world;

on a purely technical level, a very advanced artificial lighting system and the innovation represented by the scenic towers.

3.2 Vitruvius as a Source of Modern Constructive Pragmatism

Fig. 3.3 Vincenzo Scamozzi—Teatro all'Antica in Sabbioneta—interior [photo by the author]

Of particular importance are also:

the absence of both the scaenae frons and the scenic arch; Inigo Jones, in accordance with the Palladian intuition for the Olympic, had instead obtained it from the enlargement of the royal door;

the introduction of the U-shaped plan that responded both to the practical need for use as a horsewoman and as a ballroom; this aspect was resumed in the Farnese Theatre, in which the stage arch also appears for the first time in modern terms.

The death of Vespasiano Gonzaga in 1591 led to a progressive decline, followed by the plague of 1630; a recovery took place only in the second half of the eighteenth century with Maria Theresa of Austria. The theatre was subjected to several alterations, in particular in 1802, following the proven risk of collapse, the roof was rebuilt with trusses, the flooring was replaced, the curtain and the archstage introduced in the meantime were preserved, the height of the stage was changed from 1.2 to 2 m to house the machinery necessary to make the scenes mobile, the high windows on the left side of the cavea were walled up, some partitions were built at the entrance for the public and a toilet, as well as the access to the loggia. In 1941, the theatre was used as a cinema. Two main restorations followed: the one in 1950 brought to light the frescoes, the one in 1957 returned the theatre very similar to that built by Scamozzi, eliminating all the improper additions and in fact restoring all the changes described above. However, the restorations, conducted by Bruno Zevi, did not create the slope of the flooring, nor the false ceiling on the loggia; also for the roof, probably with a pavilion vault sloping down towards the back wall of the theatre, the reconstruction of the coffered ceiling was preferred.

Fig. 3.4 Vincenzo Scamozzi—the scene of Teatro all'Antica in Sabbioneta [photo by the author]

In 1964, a first reconstruction of the scene created by Tito Varisco to better understand the unity of the architectural space, an unsatisfactory solution, was replaced by the current scene created by Anna di Noto and Francesco Montuori in 1996 as part of the pilot project for the conservation of the European architectural heritage (Fig. 3.4).

By innovation, Sabbioneta constitutes the link between the Olympic Theatre in Vicenza and the Farnese Theatre in Parma. The accentuated organic vocation, also typical of the materials used, and the pursued unity of the architectural organism constitute the concrete and permanent starting point from which the subsequent phases of the architectural process descend (Fig. 3.5).

3.3 The Court Theatre of Mantua and the Farnese Theatre of Parma Towards Modern Theatre

As previously mentioned, the origins of the Court Theatre in Mantua actually constituted an antecedent and not just a product compared to the examples of Palladio and Scamozzi. In 1549, Cardinal Ercole Gonzaga commissioned Giovanni Battista Bertani, architect of the ducal factories after Giulio Romano and Girolamo Genga, the construction of a permanent theatre in the Court, located between the castle

3.3 The Court Theatre of Mantua and the Farnese Theatre ...

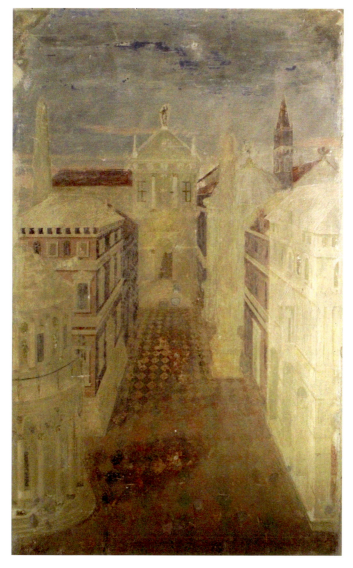

Fig. 3.5 Palazzo Giardino in Sabbioneta, Sala dei Circhi: scenographic fresco, 1588 ca. [photo by the author]

and the horsewoman. However, there is too little knowledge to be able to hypothesize a reconstruction and above all to recognize a prevalence that is not exclusively of a temporal nature over the other two models. The description of the theatre is contained in two octaves of the Edification of Mantua by Raffaello Toscano: *"Ricca è la Scena, ù gli strioni intenti/a le bell'opre concorreno spesso/i cui superbi e nobili ornamenti/mostran quant'arte l'Arte ivi abbia messo. / Di travi è fabbricata e d'assamenti/a pittura, a rilievo, e segue appresso/una Città, qual par che sia ripiena/di quant'arte e virtuti unqua ebbe Athena. / Contra il gran Palco che con gratia pende/mille gradi il Bertan pose architetto, / ch'un mezo circol fanno, e vi s'ascende/con gran facilità su fin'al tetto; / giù resta un Campo ove sovente accende/il fiero Marte a' suoi seguaci il petto; / templi, torri, palazzi e prospettive/e son, che paion vive"* Fig. 3.6. The theatre, built in a pre-existing building housing the court armory, was equipped with a semicircular cavea, with wooden steps up to the maximum reachable height delimiting the orchestra, the stage was sloping and the fixed stage composed of both plastic elements and painted elements. There was no scaenae frons, according to the paradigms introduced by Serlio and in line with the perspective scenes by Genga and Peruzzi, of which Bertani was familiar with the achievements of Urbino and Rome. The theatre, which was open to the public during Carnival, was destroyed by a fire in 1588, even before the construction of the Teatro all'Antica and of which it may have represented an antecedent, according to the similarities contained in the description. From the rebuilding that followed, based on a design by Ippolito Andreasi, therefore from 1591, the building has survived until today, although with changed functions. After that of Andreasi, the project by Antonio Maria Viani followed, equipped with tiers and for the first time with boxes which, also arranged laterally at the same exedra tiers, to occupy the space once intended for versurae, housed orchestra and gentlemen; the scene was already modern, perspective and painted on canvas. According to Tamassia,[6] which incorporates the theories of Marani, the Viani Theatre could have coexisted and did not represent a remake of that of Andreasi, smaller, corresponding in size to the Bertani Theatre, located towards piazza Sordello. The Viani Theatre, on the other hand, would have inaugurated a new tradition, continued by Bibiena and Piermarini, occupying a larger dimension of the building towards the inside.

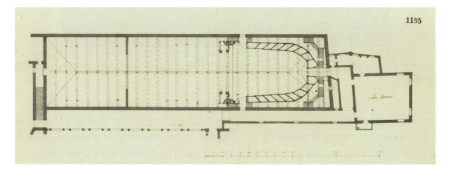

Fig. 3.6 Rilievo di Giuseppe Bianchi del Teatro Ducale, pianta [Archivio di Stato di Mantova, Archivio Gonzaga, H. VIII.—Teatri, giuochi, feste, maschere e lotterie, Teatri 1669—1787, busta n. 3170 bis]

3.3 The Court Theatre of Mantua and the Farnese Theatre ...

Proof of what was hypothesized would be the exceptionally large stage and backstage, the ancient seat of the minor theatre which, placed at a higher altitude, served as a perspective telescope if necessary. Of the theatre designed by Ferdinando Bibiena between 1706 and 1707 and completed only in 1732 by his pupil Andrea Galluzzi, it remains a survey, reproduced in Fig. 3.6, by the engineer Giuseppe Bianchi, preparatory to the reconstruction of the trussed roof which was necessary in 1775. The Bibiena Theatre was subjected to a fire in May 1781, which led to the new Piermarini project, completed just 2 years later with the construction management of Pozzo. A further survey of the Piermarini Theatre is preserved, reproduced in the following images, and some drawings showing the demolitions and reconstructions necessary to convert the building into the future Cocoon Market. These drawings, created by Carlo Andreani and still partially unpublished today (Figs. 3.7, 3.8 and 3.9), confirm the existence of the double stage. Thanks to the 1789 Description of the Regio Ducal Teatro Nuovo, we also have confirmation that it was made up as follows: a dressing room at the entrance, a ticket office, a distribution corridor, the hall of the foyer, a further room that gave access to the theatre itself, bell-shaped,

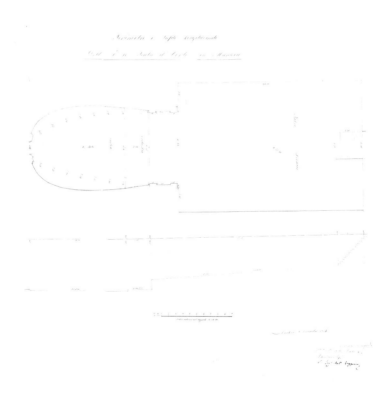

Fig. 3.7 Mappa dell'Imperial Regio Teatro di Corte in Mantova [Archivio di Stato di Mantova, estratta da: I.R. Delegazione, a. 1863, busta n. 3588]

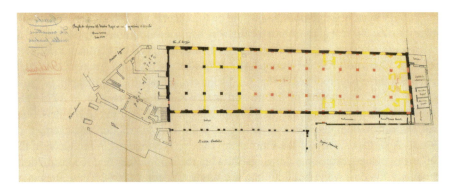

Fig. 3.8 Ex Teatro Regio, plan with demolitions and recontructions [Archivio Storico Comunale di Mantova, categoria V, classe 3, articolo 1, busta n. 30]

52 m long in total, equipped with five tiers of boxes, stage, understage and backstage, further rooms, dressing rooms and room for clothing. This conformation was not substantially dissimilar from the Bibiena project and from the previous one by Viani, which were actually taken up in full in the system. An alternative project by the young architect Antonio Colonna was presented, but not realized, and is kept in the Municipal Library of Foligno together with the original project by Piermarini. This proposed the overturning of the stalls and the stage for a more functional but expensive distribution.[7]

Ultimately, what is most striking is the suggestion that the Viani Theatre may have represented the first modern stable theatre with boxes, starting from the models of Bertani and Scamozzi, and that in turn this theatre has in fact conditioned the layout of two bell theatres now completely modern. The latter, in particular, constituted, in turn and with certainty, the model for the construction of the Teatro alla Scala in Milan.

The fact that the Mantuan Court Theatre today constitutes neither a direct source, as it no longer exists (Figs. 3.10 and 3.11), nor a complete indirect source, given the scarcity of the documents received, softens its innovative capacity, of updating the type, in favor of the subsequent Farnese Theatre.

It has been said that the Teatro all'Antica di Sabbioneta represents the link between the Olympic and the Farnese. Even the figure of Aleotti, after a long silence from critics over the past centuries, can be recognized as close to those of Palladio and Scamozzi in the process of developing the modern space for the show. Aleotti had proved his skills in the field with the Theatre of the Accademia degli Intrepidi, built in Ferrara and which was written about in the previous chapter, when Duke Ranuccio I commissioned the Farnese from him.

Manfredo Tafuri in *The Theatrical Place from Humanism to Today*[8] writes about the Farnese that "*hall and stage are architecturally unified, as an overturning of the*

3.3 The Court Theatre of Mantua and the Farnese Theatre …

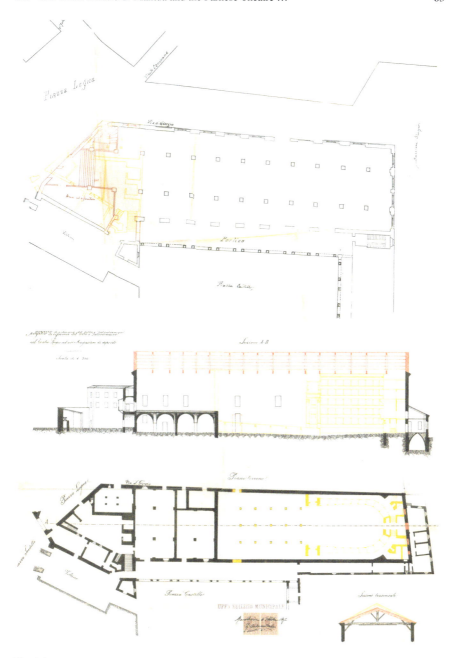

Fig. 3.9 Ex Teatro Regio, section and plans with demolitions and reconstructions [Archivio Storico Comunale di Mantova, categoria V, classe 3, articolo 1, busta n. 30]

Fig. 3.10 Palazzo Ducale in Mantua—Piazza Castello

Fig. 3.11 Palazzo Ducale in Mantua—Cortile della cavallerizza and Museo Archeologico Nazionale di Mantova [photo by the author]

urban space within the theatrical space […] and the city is no longer just the ultimate element of the spectacular vision, but it becomes theatrical space in an absolute sense". Although, like the previous Teatro all'Antica and Court Theatre, it is not an independent building, but the transformation of a large arms room arranged as a bridge between two open courtyards inside the Palazzo della Pilotta, the planimetric conformation of the serving paths on three of the four sides of the hall and the roof make it basically a served, carried, a nodal space. The hall was still incomplete in the decorative apparatus when it became a theatre, probably there was already originally a balcony at an intermediate level, it was therefore not a simple armory, but a hierarchically preponderant room, with a spectacular vocation, to the interior of the

3.3 The Court Theatre of Mantua and the Farnese Theatre …

large service building by Gerolamo Rainaldi, intended for games and entertainment, then under construction and remained unfinished. The plan dimensions are 87.22 × 32.16 m with a height of 22.67 m; according to the calculations of Carlo Donati it could reach 4350 spectators; there are 14 steps, once interrupted by the open stage for the princes, and still today surmounted by two orders of loggias. The plant has a U-shape, open to a real theatre square, very large. An innovative character covers the stage: it is equipped with a proscenium with two symmetrical projections projected towards the versurae; the giant order allows the bleachers to be welded to the stage, unifying the organism, interrupted precisely by the versurae according to the will of the client, who recognized in these real triumphal arches for the entrance of the Lords; the stage is always equipped with the first proscenium that has come up to date, to enclose a modern, illusionistic, perspective, mobile scene; the stage also has both the attic, the under and the backstage. The loggias also form a proscenium, in fact wall paintings run along the entire perimeter of the room beyond them, and the painting along the ceiling recreated the illusion of two further galleries as well as the sky.

The aforementioned ephemeral nature of the court halls and city tournaments, the specific occasion for which it was created and the haste contributed to the realization with some differences with respect to the graphic design and the wooden model.

The Farnese, which at the time was defined as the eighth wonder, was inaugurated only in 1628 with the *Mercurio and Marte* tournament by Claudio Achillini, music by Claudio Monteverdi; however it had been completed since 1619 for the entry of Cosimo II de' Medici, who then gave up the visit.[9] Alfonso Pozzo classified his *The Defense of Beauty*, a work with which the theatre was to be inaugurated, as a theatrical party. "*On the feast of 1618 the elements that made up the organic scheme of these chivalrous parties will be reworked and perfected: the knights' combat will be transformed into equestrian ballet, the recitative actions will become a dramatic representation, the provisional structure of the tournament field will finally find a location stable in the birth of a permanent theatrical architecture*" Ciancarelli reports.[10] The theatre would have constituted a documentary testimony of the reconciliation with the de' Medici family, theatre as an architectural synthesis of the festival.

The Farnese was opened a few times during the seventeenth century due to its size, in fact a second alternative theatre was built by Lolli, with a capacity of 1200 seats, in the reserve building. There was also a subsequent redevelopment project by Ferdinando Bibiena, which was never carried out. The latest events that have transported it to date are of little importance for typological purposes, as they do not represent real changes, however the reconstruction of the trusses in 1867 on a project by Engineer Mazzucchelli resulted in the definitive loss of the painted ceiling. After a happy interlude in which the theatre experienced a new use between the two wars, the Allied bombing of '44 marked the near destruction of the complex; only a careful and slow restoration that lasted a total of three decades, from 1956 to the 1980s, prevented its definitive loss.

3.4 Summary

Three prominent cases were analyzed, the Olympic Theatre in Vicenza, the Teatro all'Antica in Sabbioneta and the Farnese Theatre in Parma, all of which can be defined as *inventiones*, original contributions to the developing of the type.

The autonomy of the theatre building is fully achieved in them, they have specific permanent characteristics, in particular it has been emphasized how the Palladian versurae involve the union between the scene and the audience, anticipating the nodality of the modern theatre, towards the unification of the architectural organism, and how the set of characters of the Sabbioneta Theatre, the link between the Olympic and the Farnese, determine an accentuated organic and unity vocation.

The authors' conscious deviation from the Vitruvian model was therefore highlighted, their critical conscience, their grafted paths, contextualized in the urban fabric.

The suggestive story of the Court Theatre of Mantua was also reported on the sidelines, providing interpretative hypotheses that would see this theatre very close in character and intuition to the three most famous models.

3.4 Summary

TEATRO OLIMPICO CODE3.01

Andrea Palladio - Model for Teatro Olimpico in Vicenza - 1579 - 1580

TYPOLOGICAL SCHEME

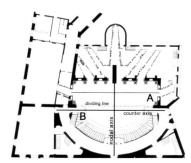

LEGEND

A STAGE
B AUDIENCE

SUMMARY OF THE MAIN SELECTION CHARACTERS

place_**VICENZA**
level of the event_**international**
site of architectural interest_**Yes**
UNESCO site_**Yes**
year_**1585**
building_**permanent theatre**
construction_**wooden and brick**
features_**stage distinct from the area dedicated to spectators**
keyword_**organism**

TEATRO DI SABBIONETA — CODE3.02

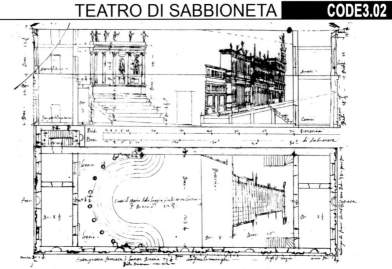

Vincenzo Scamozzi - Preparatory sketch for the Teatro di Sabbioneta - 1588

TYPOLOGICAL SCHEME

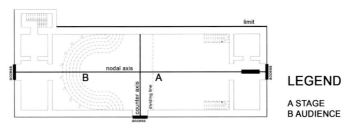

LEGEND

A STAGE
B AUDIENCE

SUMMARY OF THE MAIN SELECTION CHARACTERS

place_**SABBIONETA (MN)**
level of the event_**local**
site of architectural interest_**Yes**
UNESCO site_**Yes**
year_**1590**
building_**permanent theatre**
construction_**wooden and brick**
features_**stage distinct from the area dedicatated to spectators**
keyword_**organism**

3.4 Summary

TEATRO DI SABBIONETA — CODE3.12

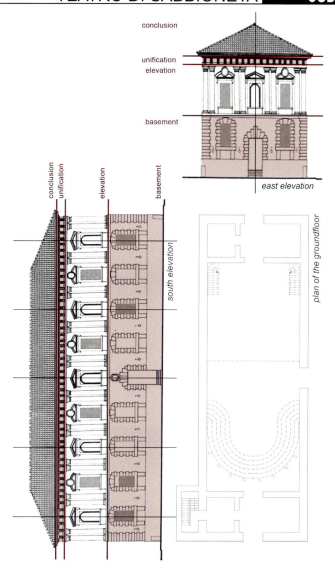

TEATRO FARNESE DI PARMA | CODE3.03

Giovan Battista Aleotti - Teatro Farnese di Parma - 2015

TYPOLOGICAL SCHEME

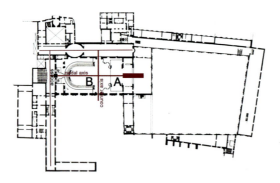

LEGEND

A STAGE
B AUDIENCE

SUMMARY OF THE MAIN SELECTION CHARACTERS

place_**PARMA**
level of the event_**international**
site of architectural interest_**Yes**
UNESCO site_**No**
year_**1628**
building_**permament theatre**
construction_**wooden and brick**
features_**proscenium**
keyword_**organism**

3.4 Summary

COMPARATIVE TABLE CODE3.24

Category	TEATRO OLIMPICO	TEATRO DI SABBIONETA	TEATRO FARNESE DI PARMA
stage	■ flat □ inclined	□ flat ■ inclined	□ flat ■ inclined
scene	■ fixed □ mobile ■ material □ perspective	■ fixed □ mobile ■ material □ perspective	□ fixed ■ mobile □ material ■ perspective
scaenae frons	■ present □ absent	□ present ■ absent	□ present ■ absent
proscenium	■ present □ absent	□ present ■ absent	■ present □ absent
separation audience/stage	□ present ■ absent	■ present □ absent	■ present □ absent
audience	□ "U" shape ■ "V" shape □ jagged □ bell shape □ circular □ oval □ elliptical □ double ellipse □ horseshoe □ mirrored □ flat ■ inclined	■ "U" shape □ "V" shape □ jagged □ bell shape □ circular □ oval □ elliptical □ double ellipse □ horseshoe □ mirrored □ flat ■ inclined	■ "U" shape □ "V" shape □ jagged □ bell shape □ circular □ oval □ elliptical □ double ellipse □ horseshoe □ mirrored □ flat ■ inclined
orchestra	■ bounded □ not bounded ■ equal □ lower	■ bounded □ not bounded ■ equal □ lower	■ bounded □ not bounded ■ equal □ lower
theatre square	■ present □ absent	■ present □ absent	■ present □ absent
seats	□ with steps ■ with steps and loggias □ with steps and boxes □ with boxes	□ with steps ■ with steps and loggias □ with steps and boxes □ with boxes	□ with steps ■ with steps and loggias □ with steps and boxes □ with boxes

COMPARATIVE TABLE — CODE3.25

Teatro di Corte di Mantova — VIANI design

Category	Options
stage	☐ flat · ■ inclined
scene	■ fixed · ☐ mobile · ☐ material · ■ perspective
scaenae frons	■ present · ☐ absent
proscenium	■ present · ☐ absent
separation audience/stage	■ present · ☐ absent
audience	☐ "U" shape · ☐ "V" shape · ☐ jagged · ■ bell shape · ☐ circular · ☐ oval · ☐ elliptical · ☐ double ellipse · ☐ horseshoe · ☐ mirrored · ☐ flat · ■ inclined
orchestra	■ bounded · ☐ not bounded · ■ equal · ☐ lower
theatre square	■ present · ☐ absent
seats	☐ with steps · ☐ with steps and loggias · ■ with steps and boxes · ☐ with boxes

Teatro di Corte di Mantova — BIBIENA design

Category	Options
stage	☐ flat · ■ inclined
scene	■ fixed · ☐ mobile · ☐ material · ■ perspective
scaenae frons	■ present · ☐ absent
proscenium	■ present · ☐ absent
separation audience/stage	■ present · ☐ absent
audience	☐ "U" shape · ☐ "V" shape · ☐ jagged · ■ bell shape · ☐ circular · ☐ oval · ☐ elliptical · ☐ double ellipse · ☐ horseshoe · ☐ mirrored · ☐ flat · ■ inclined
orchestra	■ bounded · ☐ not bounded · ☐ equal · ■ lower
theatre square	■ present · ☐ absent
seats	☐ with steps · ☐ with steps and loggias · ☐ with steps and boxes · ■ with boxes

Teatro di Corte di Mantova — PIERMARINI design

Category	Options
stage	☐ flat · ■ inclined
scene	■ fixed · ☐ mobile · ☐ material · ■ perspective
scaenae frons	■ present · ☐ absent
proscenium	■ present · ☐ absent
separation audience/stage	■ present · ☐ absent
audience	☐ "U" shape · ■ "V" shape · ☐ jagged · ■ bell shape · ☐ circular · ☐ oval · ☐ elliptical · ☐ double ellipse · ☐ horseshoe · ☐ mirrored · ☐ flat · ■ inclined
orchestra	■ bounded · ☐ not bounded · ☐ equal · ■ lower
theatre square	■ present · ☐ absent
seats	☐ with steps · ☐ with steps and loggias · ☐ with steps and boxes · ■ with boxes

Notes

1. In (1992) Il teatro Olimpico. Electa, Milano.
2. *«la volontà del Palladio di adeguare le strutture antiche a un ambiente chiuso, nato per soddisfare esigenze funzionali moderne, nel rispetto della sintesi geometrico-spaziale della cavea, dell'orchestra, del proscenio e della frons scaenae del teatro romano».* Ibidem: 69.

3.4 Summary

3. *«la novità scaturisce dal riproporre un linguaggio organicamente rivissuto, per esprimere concezioni ed esigenze moderne dell'architettura, contaminando tra di loro esperienze apparentemente contraddittorie».* Ibidem: 77.

4. *«in tal guisa esso assume il ruolo stesso di sintesi simbolica di quella che era stata la città realisticamente portata in scena dalle prospettive del teatro peruzziano-serliano. E' un nuovo modo di visualizzare lo scenario costante del teatro cinquecentesco, che è l'idea di città; in Palladio concentrata e assorbita nell'incombente columnatio, cui si uniformano tutti gli elementi architettonici, anche quelli di secondo piano».* Ibidem: 45.

5. The drawing of the theatre of Sabbioneta, consisting of the plan and the longitudinal section, is a sketch, a draft that Scamozzi made on the spot between May, 3 and 10, 1588, during his first visit to the city; the executive project has instead been lost. In October 1589, the outside of the theatre was already completed, from that date and until February 1590 Vespasiano Gonzaga personally witnessed the completion of the works together with Scamozzi.

6. Tamassia LO (1999) L'edificio del Mercato dei Bozzoli, la storia del sito. In: Quaderni di Archeologia del Mantovano Museo Civico Archeologico di Ostiglia Gruppo Archeologico Ostigliese, (1): 196.

7. Finally, from unpublished documents of the Archivio Storico Comunale of Mantua, it is deduced that in January 1822 a tender was held to reduce the square and the arcades of the Regio Teatro to a good state, destined for the newly established grain market, as well as for create five rooms for grain storage and a dedicated office. The theatre was then demolished in 1898 to allow the construction of the Bozzoli Market, completed in 1903. In 1978, the National Archaeological Museum of Mantua was established there, for the story of which see the essay *The building of the Bozzoli Market, the history of the site*, of which a significant part is reported in the apparatuses.

8. *«sala e palcoscenico sono unificati architettonicamente, come ribaltamento dello spazio urbano all'interno dello spazio teatrale […] e la città non è più solo elemento ultimo della visione spettacolare, ma si fa spazio teatrale in senso assoluto».* Essay included in (1976) Teatri e scenografie. TCI, Milano: 31.

9. In March 1618, Aleotti left Parma, however, proving to be inclined to return for the last few needs before an imminent inauguration should it occur, leaving Alfonso Pozzo to direct the construction site, as well as Enzo Bentivoglio to perform the functions of impresario. The Bentivoglio family after the Este devolution had represented his main client for Aleotti.

10. *«Nella festa del 1618 gli elementi che costituivano l'organico schema di queste feste cavalleresche verranno rielaborati e perfezionati: il combattimento dei cavalieri si trasformerà in balletto equestre, le azioni recitative diverranno rappresentazione drammatica, la struttura provvisoria del campo del torneo troverà infine una sede stabile nella nascita di un'architettura teatrale permanente».* In (1987) Il progetto di una festa barocca. Bulzoni Editore, Roma.

References

(1989) Giulio Romano. Electa, Milano
(1989) Giulio Romano. Accademia Nazionale Virgiliana, Mantova
(1991) Il Teatro all'Antica di Sabbioneta. Il Bulino edizioni d'arte, Modena
Algeri G (2003) Il Palazzo Ducale di Mantova. Editoriale Sometti, Mantova
Bazzotti U (2012) Palazzo Te a Mantova. Franco Cosimo Panini, Modena
Del Valle Ojeda M, Presotto M (2013) Teatro clasico italiano y espanol. Sabbioneta y los lugares del teatro. PUV, Valencia
Magagnato L (1992) Il teatro Olimpico. Milano
Mazzoni S, Guaita O (1985) Il teatro di Sabbioneta. Leo S. Olschki Editore, Città di Castello
Mazzoni S (1992) Il teatro Olimpico di Vicenza e l'arte della memoria. Medioevo e Rinascimento VI(III):97–114
Millette D (2016) Vitruvius and the Re-invention of classical theatre architecture. In: Sanvito P (ed) Vitruvianism. Origins and transformations. De Gruyter, Berlino
Plate Tschudi V (2008) Serlio and Sabbioneta. A city built of prints. In Eriksen R (ed) Rethoric, Theatre and the Arts of Design. Novus Press, Oslo
Sanvito P (2016) How much Vitruvianism is left in Vincenzo Scamozzi's Architectural Theory? Vitruvius as a source of early modern pragmatism. In: Sanvito P (ed) Vitruvianism. Origins and Transformations. De Gruyter, Berlino
Strappa G (1995) Unità dell'organismo architettonico. Edizioni Dedalo, Bari
Tamassia LO (1999) L'edificio del mercato dei bozzoli. La storia del sito. In: Quaderni di Archeologia del Mantovano (1):191–223

Documents

Giovanni Battista Aleotti, Pianta del Teatro del serenissimo Duca di Parma del salone grande fatto da ingegnero Argenta [1617–1619], 119x530mm [Biblioteca Comunale Ariostea, Raccolta Aleotti 162]
Giovanni Battista Aleotti, Alzato del Teatro Farnese di Parma. Progetto non realizzato [1615–1617], 550x335mm [Biblioteca Comunale Ariostea, Raccolta Aleotti 165]
Teatri 1669–1787 [Archivio Gonzaga, H. VIII.—Teatri, giuochi, feste, maschere e lotterie, busta n. 3170 bis]
Suddivisione 37—piazza Castello, Fascicolo n. 1 [Archivio Storico Comunale di Mantova, Titolo XI—Lavori Pubblici, Articolo 21° Strade Comunali Urbane, 1819–1851, busta n. 682]
Mappa dell'Imperial Regio Teatro di Corte in Mantova [Archivio di Stato di Mantova, estratta da: I.R. Delegazione, a. 1863, busta n. 3588]
Ex Teatro Regio, Mercato Frutta e Verdura [Archivio Storico Comunale di Mantova, categoria V, classe 3, articolo 1, busta n. 30]
Teatro Regio [Archivio Storico Comunale di Mantova, Categoria V—Finanze, Contabilità, Patrimonio e Beni Demaniali, Classe 3 Patrimonio e Beni di Demanio Comunale, Articolo 1 Edifici, Stabili ed altre proprietà, 1901–1938]

Chapter 4
Theatre as Recast

The Process Described Starting from the History of Mantua

Abstract The chapter starts from some particularly significant events in theatres in Mantua, from treatises and from contemporary examples such as the Teatro della Compagnia by Natalini, to describe the developing process of the modern space for the show by recasting, in pieces of cities far from the exceptional, so far analyzed, of the court and the academy.

Keywords Recast · Mantua · Academy · Giovan Battista Aleotti

4.1 Old and New Theatres in Mantua

Beyond the aforementioned Court Theatre and the famous example of the Bibiena Scientific Theatre, analyzed below, we will focus in the following on some case studies subject to heavy alterations over time and of which particularly close and unique is the relationship with basic buildings in general: they originated from them and returned to them.

Therefore, through original documents in part unpublished, the phases of the developing process by recasting of some of the major theatres no longer existing and linked to popular representations are reconstructed: the Teatro Vecchio, the Fedeli Theatre and the Arnoldi Theatre. As already mentioned and also confirmed by the contemporary treatise by Carini Motta, theatres move away from the courts and academies, which nevertheless remain models of reference, to be reborn, public, starting from new urban contexts.

In Urbis Mantuae Descriptio of 1628, Bertazzolo (Fig. 4.1) records the presence of a public stage, between the current via del Teatro Vecchio and piazza Arche, further to the Court Theatre. Even the chronicler Federigo Amadei reports the existence of a covered passage that put the Ducal Palace in direct communication with the theatre, which was therefore not within the prince's city, but at the court's disposal and also open to the public for a fee. Indeed, as well as being a public scene, it was also defined as a small scene to distinguish it from the Court Theatre. In all likelihood, the time of its construction is earlier, coinciding with the born of an art company under the control of Duke Vincenzo. From archival documents dating back to 1609, we learn that the "comedian's room" was equipped with boxes, which the Duke gave

© Springer Nature Switzerland AG 2022
S. Clemente, *Reconstructing Theatre Architecture*, The Urban Book Series,
https://doi.org/10.1007/978-3-030-89968-4_4

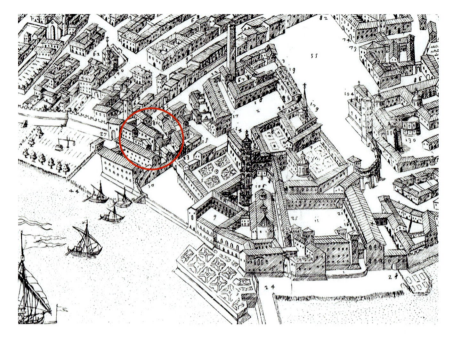

Fig. 4.1 Detail of the Urbis Mauntuae Descriptio by Bertazzolo (1628) with Teatro Vecchio highlighted [elaboration by the author]

to comedians during the carnival period. At the beginning of the seventeenth century, therefore, the Gonzagas had two theatres: one, the most important, located in the heart of the palace in the form of a city, reserved for the cultured spectacle of the court; the other, located on the edge of the prince's palace and intended for more popular performances, also open to a different public in terms of wealth and social rank.[1] It is difficult to hypothesize, apart from the described presence of boxes, the morphology of the theatre in this phase, given the scarcity of sources. It was probably Viani who built the first building of the theatre, also known as the Teatro dei Comici. In 1688, Fabrizio Carini Motta, who in 1671 had obtained the title of "superintendent of the factories and prefect of theatres", carried out a project for the adaptation of the Teatro Vecchio, contained in the 1767 survey by Giovanni Cadioli. From the comparison between *Pianta del Teatro Vecchio e progetto per annettervi una casa contigua in cui ricavare camerini di retropalco* by Cadioli (Fig. 4.2) and the previous *Spiegazione della Pianta del Teatro Vecchio con la Distinta delle Case contigue di ragione della Regio Ducal Camera di Mantova, che sarebbero necessarie d'aggregarsi al detto Teatro per li giuochi alfine di abilitare il detto Teatro a ricavarne un congruo prodotto* (Fig. 4.3), in 1755, both at Archivio di Stato in Mantua, it is possible to appreciate the innovations introduced by Motta and Cadioli and to reconstruct the last phases of the theatre building before the French siege, which forced the Austrians, in 1797, to make the theatre firewood.

4.1 Old and New Theatres in Mantua

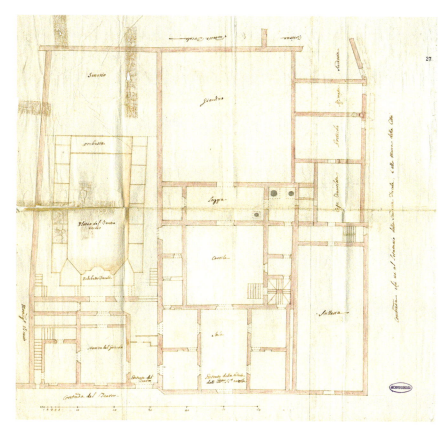

Fig. 4.2 Spiegazione della Pianta del Teatro Vecchio con la Distinta delle Case contigue di ragione della Regio Ducal Camera di Mantova, che sarebbero necessarie d'aggregarsi al detto Teatro per li giuochi alfine di abilitare il detto Teatro a ricavarne un congruo prodotto [Archivio di Stato di Mantova, Archivio Gonzaga, H. VIII. Teatri, giuochi, feste, maschere e lotterie, busta n. 3170]

The first configuration returns a theatre constituting a portion of a more complex building, articulated around a central courtyard and a larger internal garden. The boxes are arranged orthogonally to each other, with the ducal box set against the orchestra and the stage. The stalls are arranged centrally, in the absence of steps. From the Description of the Teatro Vecchio in Mantua dated March 26, 1755, it results in particular: an entrance with a ticket office; the stalls, with wooden floor and 22 rows of seats, 11 on each side, for a total of 130 seats; the orchestra with parapet, which was accessed through a door located on the right looking at the stage; the understage, which was entered through two small doors placed in turn in the orchestra and which featured 3 bridges and 30 guides to move the scenes; the stage, which was accessed via an additional ladder in the orchestra, illuminated by large windows and equipped with 13 cuts on each side; two stairways led up to the attic, the boxes were 16 on the ground floor plus the ducal box, 18 in the first order, 20 in

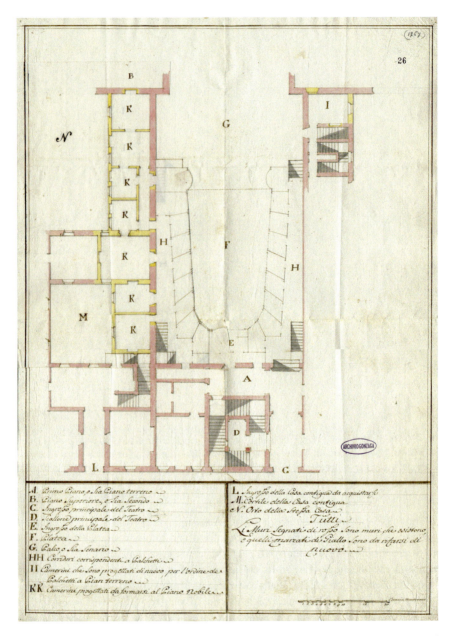

Fig. 4.3 Teatro Vecchio [Archivio di Stato di Mantova, Archivio Gonzaga, H. VIII. Teatri, giuochi, feste, maschere e lotterie, busta n. 3170]

4.1 Old and New Theatres in Mantua

the second, third and fourth, the fifth order instead presented 10 since the remaining space was made available to servants; the theatre was finally illuminated by a six-sided luminary, each equipped with a torch. In the extension of Cadioli, there is the appearance of the bell shape, to optimize the number and quality of the observation points of the private boxes, the definition of a system of accesses and horizontal and vertical distribution paths that give greater autonomy to the theatre building; in particular, it is assumed that the connecting stairs to the attic will be moved outside the area occupied by the stage, given its small size. The extension of Cadioli transforms the access path to the boxes into a path that is also a distribution path for the rooms to be used on the various floors with a noble reduced, a small hall and dressing rooms serving the theatre, previously constituting the rooms of an adjoining house; this path can therefore be understood as an overturning of the external one. In fact, the theatre acquires, by recasting, a second autonomous front, further to the one on which the entrance stands, and is taking shape as a nodal space, even if still only partially and incomplete.

Of the Fedeli Theatre, built by Fabrizio Carini Motta probably before 1669, located in the district of the Nave, behind the port of the Catena and precisely in via Cardone, a heavily remodeled building remains today, of which only the facade on the alley of Canove has partially preserved its ancient splendor (Fig. 4.4). In fact, from archive documents dating back to 1669, it appears that the theatre of comedies "*with the rooms attached to both sides, however much the width of the theatre holds, and not beyond, and no longer beyond*", would have been, on the death of the owners, transferred to Duke Ferdinando Carlo and his successors. The theatre was of the

Fig. 4.4 Place where Teatro Fedeli stood [photo by the author]

type with boxes, distributed over five orders, it also presented the ducal box and the stage with fixed houses to constitute the scene. The paternity of the theatre, confirmed in the *Treatise on the structure of the theatre, and scenes*, earned Carini Motta a visiting card at the Gonzaga court, where he finally acquired the title of ducal architect in 1672. This information, beyond the prospectus, and the manifest reconversion and re-understanding of the building into a purely residential fabric, are however insufficient to reconstruct the phases up to now; the last testimony relating to it before the changes that affected it dates back to 7 years after the death of the owner, when the theatre was still working.

Finally, from unpublished documents of the Archivio Storico Comunale in Mantua reproduced in Figs. 4.5, 4.6, 4.7, 4.8, 4.9 and 4.10, we finally obtain an episode subsequent to those described and yet very similar to them. The Chief Master Arnoldi asked permission from the Municipal Council of Mantua to change the elevation of some residential buildings, adjacent to the Mainolda prison, which were then transformed into a theatre. The request for the permit, dated June 25, 1882, was followed by the favorable opinion communicated to the applicant on July 2, containing graphic indications for the modification of the openings on the main façade of the theatre. On June 14, 1883, Arnoldi was working, as evidenced by a new letter, to obtain the surveillance of the fire brigade and regulate the flow of carriages and pedestrians entering and leaving during the shows, because of the area of strongly compact residential fabric in which the theatre stood. The plant is precious since it clarifies its urban insertion, identifying its pedestrian and vehicular accesses. The theatre, owned by Cesare Bonoris, was illuminated laterally by large windows, had a horseshoe-shaped room, equipped with four tiers of boxes with loggias inserted in the first and third tier, wood finishes, Moorish-style proscenium; the roof consisted of a flat ceiling with a rectangular movable curtain; there were also two foyers for smokers and two coffee rooms. From the testimonies, we know that the contemporaries nevertheless found the stalls not very spacious, the gallery very uncomfortable, and that the theatre had serious problems of acoustic insulation: the public in fact gathered in the alleys and listened to the show perfectly, without entering it. The Arnoldi Theatre is a unique testimony of its kind in that it is the product, by recasting, of several contiguous terraced houses, intended for use to which the environments have naturally returned, reversing the transformation process typical of the urban fabric (Fig. 4.11).

4.2 Wonderful Anticipations and Surprising Delays by an Author/Reader: Fabrizio Carini Motta

The attention is therefore placed on the rediscovery in recent times of a text of 1676, *Trattato sopra la struttura de' theatri, e scene, che à nostri giorni si costumano, e delle Regole per far quelli con proporzione secondo l'insegnamento della pratica Maestra Comune* by Fabricio Carini Motta, Architect of the Most Serene Duke of Mantua, to whom the described examples of modern theatre formation by recasting

4.2 Wonderful Anticipations and Surprising Delays by an Author … 81

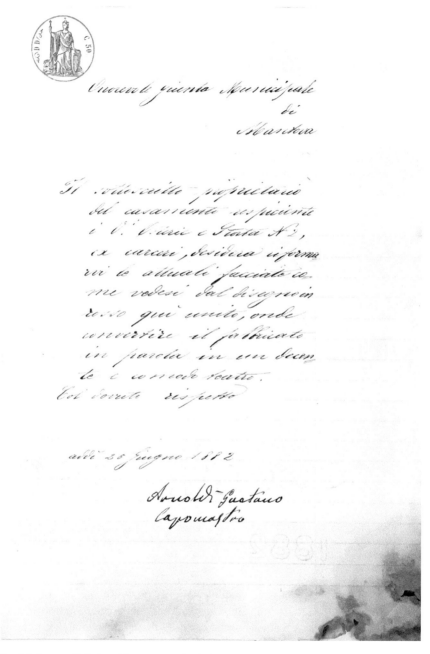

Fig. 4.5 Lettera di G. Arnoldi contenente la richiesta prodotta alla Giunta Municipale di Mantova di variazione dei prospetti per la realizzazione del Teatro Arnoldi [Archivio Storico Comunale di Mantova, Titolo XIV—Sicurezza Pubblica, Articolo 24° Teatri, 1882–1884, Suddivisione 6—Teatro Arnoldi, Fascicolo n.1, busta n. 1093]

Fig. 4.6 Stato di fatto del futuro Teatro Arnoldi—prospetti lungo via Osterie e via Storta [Archivio Storico Comunale di Mantova, Titolo XIV—Sicurezza Pubblica, Articolo 24° Teatri, 1882–1884, Suddivisione 6—Teatro Arnoldi, Fascicolo n.1, busta n. 1093]

are closely linked. The treaty had multiple fortunes: that of being the only one of its kind to come down to us and that of being, if the combination is allowed, a *De architectura* of the seventeenth-century theatre, all the more precious the rarer the tangible structures to which it refers remained standing. Designing as a synonym for reading and rereading.

The treatise unjustly neglected in the past, essentially because it was printed in just over 50 copies and therefore known almost only to the court of Mantua, already in the title contains two elements of absolute novelty: the theatres as buildings are the

4.2 Wonderful Anticipations and Surprising Delays by an Author …

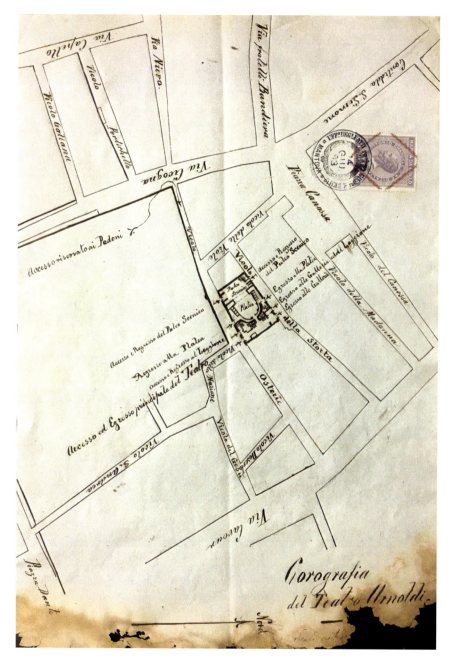

Fig. 4.7 Corografia del Teatro Arnoldi [Archivio Storico Comunale di Mantova, Titolo XIV—Sicurezza Pubblica, Articolo 24° Teatri, 1882–1884, Suddivisione 6—Teatro Arnoldi, Fascicolo n.1, busta n. 1093]

Fig. 4.8 Progetto del Teatro Arnoldi—prospetti lungo via Osterie e via Storta [Archivio Storico Comunale di Mantova, Titolo XIV—Sicurezza Pubblica, Articolo 24° Teatri, 1882–1884, Suddivisione 6—Teatro Arnoldi, Fascicolo n.1, busta n. 1093]

main object, so that it can be defined the first real treatise on theatre, and describes theatres as they costume themselves, that is, how they adapt to the time in which it was written, for the first time not referring to the classics. On the contrary, the public, popular, placed on a par with the court theatres and academies, find the dignity of description, just in support of what has been analyzed so far in the chapter, with respect to which they constitute a constructive variant.

4.2 Wonderful Anticipations and Surprising Delays by an Author …

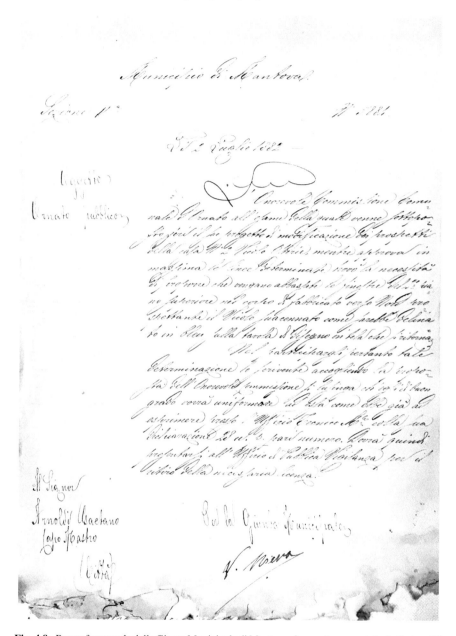

Fig. 4.9 Parere favorevole della Giunta Municipale di Mantova al progetto presentato da G. Arnoldi [Archivio Storico Comunale di Mantova, Titolo XIV—Sicurezza Pubblica, Articolo 24° Teatri, 1882–1884, Suddivisione 6—Teatro Arnoldi, Fascicolo n.1, busta n. 1093]

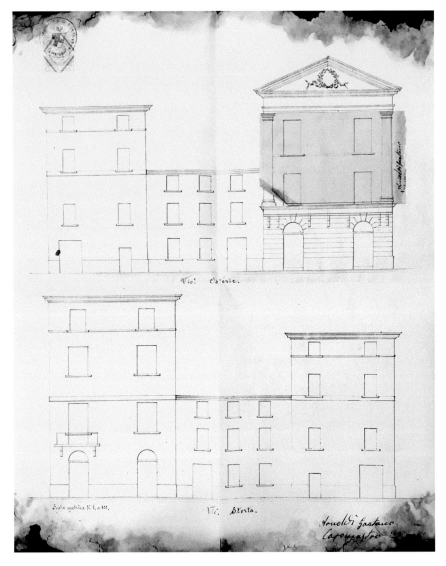

Fig. 4.10 Progetto del Teatro Arnoldi—prospetti lungo via Osterie e via Storta, contenente la variazione richiesta dalla Giunta Municipale di Mantova [Archivio Storico Comunale di Mantova, Titolo XIV—Sicurezza Pubblica, Articolo 24° Teatri, 1882–1884, Suddivisione 6—Teatro Arnoldi, Fascicolo n.1, busta n. 1093]

4.2 Wonderful Anticipations and Surprising Delays by an Author … 87

Fig. 4.11 The block today [elaboration by the author]

The treatise was published significantly before Carlo Fontana's creation of the Tordinona Theatre, which will be analyzed in the next chapter, and after the construction of the Fedeli Theatre, of which, it has been said, constitutes an important testimony. Although the treatise is backward in terms of architecture and scenographic technique, it is precious because it sheds light on little researched themes in the period between the sixteenth and seventeenth centuries. It is also underlined that up to Carini Motta we came across never specific treatises on the theatre building and instead multidisciplinary, containing in-depth analyzes relating to perspective, painting, construction materials, areas that variously and equally interested architects, painters, scenographers and mathematicians (think of the published shortly before, in 1638, *Pratica di fabricar scene e macchine ne' teatr* by Sabatini or even the subsequent and reduced to perspectives *L'Architettura civile preparata su la geometria e ridotta alle prospettive* written by Ferdinando Bibiena in 1711).

In detail, the treatise consists of 24 pages, 11 tables reproducing three different elevations and five different plans, and three further tables inserted in the text. A brief description of the contents is given below which makes it possible to identify the structuring and codified characters at the basis of the continuity of the developing process, as well as to clearly place the types identified up to then, essentially distinguishable in the two main categories of court theatre and popular theatre:

in the chapter entitled *General rules to be observed in theatres*, Carini Motta defines the concept of the theatre square, the large area occupied temporarily and partially by the stalls as well as hosting tournaments and naumachias (think of the Farnese Theatre); he is also significantly opposed to the continuity between boxes and stage for issues that appear weak to the modern eye, related to poor view and the impossibility of placing the access and exit doors between the two elements; there is also a complete definition of the proscenium;

in the next chapter, relating to the things that are necessary in the stage, he suggests a now complete understanding and codification of the modern stage, equipped with attic, understage and backstage (Fig. 4.12);

it follows an application example of the construction of the modern theatre starting from the specific modalities and proportions previously mentioned;

the chapter *Per collocare il ponto orizzontale per digradare il palco* deepens the construction of the slope of the stage and of the cavea;

in *Per far il teatro con il proscenio senza faccia, e che li fianchi siano congionti al medemo* Motta finally introduces the horseshoe plan: "*The square of that has the similarity to an imprinted shape of a horse's foot*"; he believes that this shape, suggested by Leon Battista Alberti, lacks majesty but is the best acoustically speaking; this

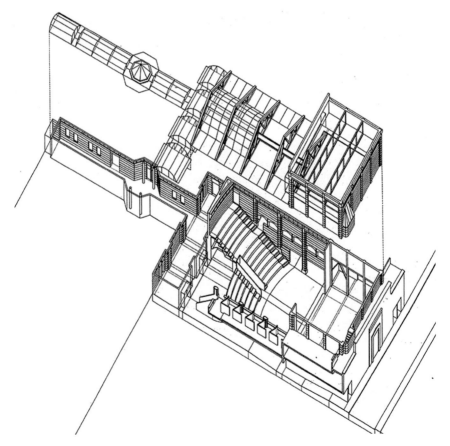

Fig. 4.12 Adolfo Natalini—Teatro della Compagnia, axonometry [http://www.nataliniarchitetti.com/progetti/spettacolo/8401.html]

constitutes an anticipation of the emergence of the type that more than any other achieves the functional, structural and stylistic unity of the architectural organism;

the chapters *Come si facciano li teatri con gradi, e con un ordine de' palchetti* e *Il teatro con palchetti come si faccia* encode the characters of the court theatre;

in *How to make theatres with divided boxes* we finally have the already mentioned description of the popular box theatre, the so-called Italian theatre in germ; Carini Motta focuses on the methods of subdivision of the boxes to ensure the best visibility and specifies the management regime of the same, real private sitting rooms from which you are free to enter and exit at will, without even being seen, along a distributive corridor, serving; it can be definitively affirmed that the developing process is unique, also started from the urban fabric, which has seen the transformation, due to the progressive specialization of the basic building, be it the form of simple living in the house, or the building, already at a next step, in fact, towards specialization;

the last chapter specifies the realization of the Fedeli Theatre according to the canons described as well as the provision to publish a second treatise on scenography, which is not analyzed in this context but of which the completely illustrated unpublished eighteenth-century copy is referred to, kept at the Biblioteca Estense in Modena.

4.3 A City Interior: The Disguised Theatre by Adolfo Natalini

The theatre by Adolfo Natalini, now used as a cinema without however having undergone significant changes from the time of construction, is a current example of theatre obtained by means of a recast. Its entrance on via Cavour, in Florence, has been defined as almost totally anonymous, unexpectedly inserted along a front of a compact block, of which it occupies a single center distance corresponding to that of a terraced house. In fact, the entrance leads into a gallery overlooked by services such as bar, ticket office, cloakroom, bathrooms; then we reach a transversal atrium covered by a barrel vault, and from here to the room, rectangular, divided into stalls and gallery or more correctly steps. This expands in the intermediate part of the block, constituting the former green areas, garden, afferent, from the beginning of the twentieth century to Bastogi Palace. This originally open portion is therefore closed and knotted, creating an infill, also facing the opposite elevation for a greater extension, corresponding to the backstage.

Internally, the theatre has explicit references to the first theatres of modernity, including the Farnese in Parma, which inspired the entrance to the hall itself. The walls are articulated by pilasters and loggias in which the boxes are obtained; the use of the Santafiora stone refers to the tragic scene (Fig. 4.13); the roof is made up of metal trusses, a reference to everyday life, to the comic scene; the fantastic bestiary made up of lamps and monsters alludes to the satirical scene; finally, the scenic tower represents the only visible external architectural emergence.

Fig. 4.13 Adolfo Natalini—Teatro della Compagnia, view of the sage, sketch of the audience [http://www.nataliniarchitetti.com/progetti/spettacolo/8401.html]

4.4 Summary

Several case studies were analyzed in the city of Mantua, describing theatre as a recast product starting from basic buildings.

Also thanks to the contribution of Carini Motta and his *Treatise on the structure of theatres and scenes*, the uniqueness of the developing process of modern theatre has been demonstrated, due to the progressive specialization of basic buildings, be it the form of living simple of the house, or of the building, already at a later level, therefore, towards the same specialization. The descriptions of the court theatre and the public theatre with boxes correspond to the two starting points, actually of equal significance, the germ of the Italian theatre, equally dignified by Carini Motta in his treatise. Finally, the birth of the theatre by recasting is always alive, as demonstrated by the contemporary example of the Teatro della Compagnia by Natalini.

TEATRO DELLA COMPAGNIA — CODE 4.01

Adolfo Natalini - Teatro della Compagnia - elevations -1987

TYPOLOGICAL SCHEME

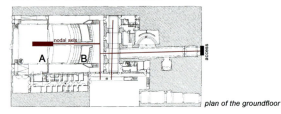

plan of the groundfloor

grafting

section

LEGEND

A STAGE
B AUDIENCE

SUMMARY OF THE MAIN SELECTION CHARACTERS

place_**FIRENZE**
level of the event_**locale**
site of architectural interest_**Yes**
UNESCO site_**Yes**
year_**1987**
building_**permament theatre**
construction_**reinforced concrete, still**
features_**stage distinct from the area dedicated to spectators**
keyword_**recast**

4.4 Summary

COMPARATIVE TABLE — CODE 4.21

TEATRO VECCHIO

- **stage:** □ flat ■ inclined
- **scene:** □ fixed ■ mobile | □ material ■ perspective
- **scaenae frons:** □ present ■ absent
- **proscenium:** ■ present □ absent
- **separation audience/stage:** □ present ■ absent
- **audience:** □ "U" shape ■ "V" shape □ jagged □ bell shape □ circular □ oval □ elliptical □ double ellipse □ horseshoe □ mirrored | □ flat ■ inclined
- **orchestra:** ■ bounded □ not bounded | □ equal ■ lower
- **theatre square:** □ present ■ absent
- **seats:** □ with steps ■ with steps and loggias □ with steps and boxes ■ with boxes

TEATRO ARNOLDI

- **stage:** □ flat ■ inclined
- **scene:** □ fixed ■ mobile | □ material ■ perspective
- **scaenae frons:** ■ present □ absent
- **proscenium:** ■ present □ absent
- **separation audience/stage:** ■ present □ absent
- **audience:** □ "U" shape ■ "V" shape □ jagged □ bell shape □ circular □ oval □ elliptical □ double ellipse □ horseshoe □ mirrored | □ flat ■ inclined
- **orchestra:** ■ bounded □ not bounded | ■ equal □ lower
- **theatre square:** ■ present □ absent
- **seats:** □ with steps □ with steps and loggias □ with steps and boxes ■ with boxes

TEATRO DELLA COMPAGNIA

- **stage:** □ flat ■ inclined
- **scene:** □ fixed ■ mobile | □ material ■ perspective
- **scaenae frons:** □ present ■ absent
- **proscenium:** ■ present □ absent
- **separation audience/stage:** ■ present □ absent
- **audience:** □ "U" shape ■ "V" shape □ jagged □ bell shape □ circular □ oval □ elliptical □ double ellipse □ horseshoe □ mirrored | □ flat ■ inclined
- **orchestra:** □ bounded □ not bounded | ■ equal □ lower
- **theatre square:** □ present ■ absent
- **seats:** □ with steps ■ with steps and loggias □ with steps and boxes □ with boxes

94   4 Theatre as Recast

Notes

1. Over the course of time, these two main theatres were joined by further ones, both temporary and occasional. On an area of 35,000 square meters of covered surface of the Ducal Palace, for which the years of Guglielmo Gonzaga until 1587 were decisive, which led to the orderly and rational complex of factories known today and the so-called palace in the form of a city, the permanent castle theatre and the temporary theatre of the Sala di Troia were built. We have evidence of the first one thanks to the plan of the noble floor outlined in 1859 by Luigi Marini; it was included between the camera picta and the hall of the initials, it also had boxes arranged in a bell shape. Built after 1630 and unused as early as 1773, at the beginning of the 1900s, it was completely demolished. Temporary performances also took place frequently in piazza Castello and in the Cortile della Cavallerizza.

References

Amadei G (1968) Un secolo su Mantova. CITEM, Mantova
Burattelli C (1999) Spettacoli di corte a Mantova tra Cinque e Seicento. Casa Editrice Le Lettere, Firenze
Ferrari D (2000) La città fortificata. Il Bulino edizioni d'arte, Modena
Natalini A (1988) Non sarà facile trovarlo. Il teatro della Compagnia a Firenze. Lotus (58):95–106
Natta F (2015) Epistolografia e teatro. Nuove considerazioni sulla scena cinquecentesca attraverso le missive della corte mantovana. In Secchi Tarugi L (ed) Pio II nell'epistolografia del Rinascimento. Franco Casati Editore, Firenze
Savi V (1983) Storia di un progetto. Adolfo Natalini e una sua architettura. Lotus (40):19–22
Strappa G (2016) Arte e scienza dei tessuti storici. U+D Urbanform and Design (3/4)
Zuccoli N (2005) Teatri storici nel territorio mantovano. Gianluigi Arcari Editore, Mantova
Duilio R, Lupano M (2015) Aldo Andreani 1887–1971 visioni, costruzioni, immagini. Electa, Roma

Documents

Mantova assediata dall'esercito Imperiale anno 1629 presa 1630, Incisione, 406x532mm [Biblioteca Comunale Ariostea, Serie XVI-134A]
Pierre Mortier dis. E inc., La Ville de Mantoue, Amsterdam, Incisione, 428x615mm [Biblioteca Comunale Ariostea, Serie XVI-134B]
TORELLI P., *L'archivio Gonzaga di Mantova*, Arnaldo Forni Editore, Bologna 1920
Teatri 1669–1787 [Archivio di Stato di Mantova, Archivio Gonzaga, H. VIII. Teatri, giuochi, feste, maschere e lotterie, busta n. 3170]
Spiegazione della Pianta del Teatro Vecchio con la Distinta delle Case contigue di ragione della Regio Ducal Camera di Mantova, che sarebbero necessarie d'aggregarsi al detto Teatro per li giuochi alfine di abilitare il detto Teatro a ricavarne un congruo prodotto

References

Suddivisione 226—via Vescovado, Fascicolo n. 1 [Archivio Storico Comunale di Mantova, Titolo XI—Lavori Pubblici, Articolo 21° Strade Comunali Urbane, 1825–1857, busta n. 730]
Suddivisione 6—Teatro Arnoldi, Fascicolo n.1 [Archivio Storico Comunale di Mantova, Titolo XIV—Sicurezza Pubblica, Articolo 24° Teatri, 1882–1884, busta n. 1093]

Chapter 5
Theatre as Urban Knot

Type Features

Abstract As we have seen, modern theatre progressively leaves the space of the court and the academy to graft once again into the urban fabric, going from representing an elite to representing larger communities. The two models coexist for a long time, until the complete codification of the Italian theatre. The latter concretizes the popular theatre in the square, revives the theatre from basic buildings. It can be said that the architectural model of the hall with overlapping loggias has essentially never been abandoned, while the planimetric conformation has been and is the subject of continuous transformations. Therefore, some examples that have made it successful throughout Europe are analyzed below. It will also be seen how the spread of the Italian theatre type is inextricably linked to that of melodrama, a real independent cultural field that is new to the previous classical and medieval traditions. It is significant, then, that a return to the city similar to that described, if not more radical, has also occurred at present in the national scene; the model of the space for the constructed, codified, standardized spectacle is rejected and new or rediscovered places are used, bounded and adapted to the function. It will therefore deal with the concept of specialization of the contemporary building organism starting from the adaptation of the pre-existing structures, using case studies such as the very well-known Palazzina Liberty. In this process, we can therefore read cyclical back and forth, with a strong generator value.

Keywords Italian theatre · Urban fabric · Urban knot · Melodrama

5.1 Public Theatre with Boxes, Italian Theatre

The *conversion of Italy to the Baroque*[1] and the advent of melodrama constituted a significant impetus for the codification of the Italian theatre type. The San Cassiano Theatre in Venice is recognized as the first to have been built for opera in 1637. However, there are very few descriptive sources of the same, including the main one that still remains the testimony of Sansovino: "*Not far from from this temple there are two beautiful theatres built at great expense, one in an ovate shape and the other round, capable of a large number of people; to recite in Carnival times, Comedies, according to the custom of the city*". The San Cassiano, with an ovate

---

© Springer Nature Switzerland AG 2022
S. Clemente, *Reconstructing Theatre Architecture*, The Urban Book Series,
https://doi.org/10.1007/978-3-030-89968-4_5

plan, is also mentioned as the old theatre, located at the bottom of the Michiela court, differently from the new or Tron theatre, which, with a round shape, stood in the disreputable area of Carampane, was rebuilt in stone and finally demolished. The short life of the San Cassiano, which did not even experience solidification, as well as the aforementioned scarce documentary sources, increase the importance of the Carini Motta treatise and the evidence still visible on the Mantuan buildings described above.

The treatise by Carini Motta, it has been said, also anticipates Carlo Fontana's project for the Tordinona Theatre (Fig. 5.1), an important example for the identification of the main characters of the Italian type. The Roman theatre is the first to bring back the horseshoe plan and to present a stage configuration suitable for movable wings, aspects that for two centuries and beyond did not undergo particular innovations, and whose climax in the quality of the execution and in the dimensions is represented by the Teatro alla Scala in Milan. Before the Tordinona Theatre, in Rome, the show was an elite phenomenon, it was placed in pre-existing buildings, it was made purely of wood without revealing its structure on the outside, therefore, it was devoid of a representative function, of the recognition of the elevation, resulting from being an independent building, an aspect anticipated instead by the theatre in Sabbioneta. The Tordinona therefore constitutes a significant innovation towards the codification of the type, even more because it was built in the context of the first public theatre, away from the court and the academy and instead absorbed in the city.

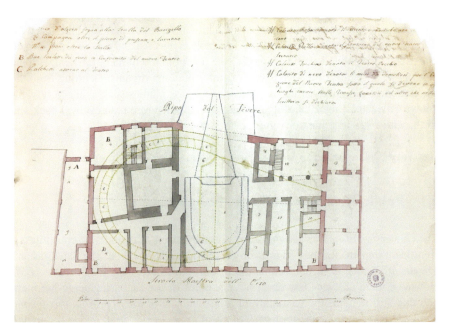

Fig. 5.1 Vecchio e nuovo teatro da costruirsi, con in evidenza le demolizioni [Archivio di Stato di Roma, Camerale III (secoli XV–XIX), busta n. 2127]

5.1 Public Theatre with Boxes, Italian Theatre

The Capranica, the della Pace, the Argentina Theatres are all subsequent examples, which mark, during the eighteenth century, the great development of public halls as permanent autonomous buildings. The first project still appears far from obtaining the autonomous knotted space of the Italian theatre. Built by Carlo Fontana on behalf of the Count d'Alibert, the favorite of Queen Christina of Sweden, it dates back to 1666. It is a rectangular hall squeezed between two buildings previously used as a prison, converted into a hotel and inn. The hall had, upon construction, an average width of 16 m and a length of 22 m, 12 boxes for each of the 6 orders arranged according to a U-shaped plan with slightly flared wings and for which no real distribution areas were specified. The main entrance was located on via Tordinona, the secondary, reserved for the nobility, took place from piazza San Salvatore in Lauro across a hanging bridge on via Tordinona, the Queen finally accessing it through an adjoining house towards the Sant'Angelo Bridge. The stage underwent a first expansion, reaching a total depth of 15 m, again based on a design by Fontana in 1671, by adding a wooden "casone" jutting out over the Tiber. This intervention constitutes the integration of a naturalistic element in the theatrical scene, the river, the fulcrum and engine of the city development. However, the successful insertion into the urban fabric as an independent building was reached almost 20 years later with the second reconstruction project of 1695, again by Fontana. As can be seen from the plan below, the orientation of the room is varied by 90°, thus marking not only a clear enlargement (the proscenium alone reaches the size corresponding to the entire room originally), but also the total occupation of the building. The theatre is therefore autonomous, and for the first time adopts the horseshoe plan, described by Carini Motta as a quote from Alberti's *De re aedificatoria*, along which 35 boxes are distributed for each of the 6 orders; the boxes adhere to the stage, to emphasize the unity of the knotted architectural organism. The theatre, one of the largest in Europe, is no longer part of either the court building or the academy, or of the block constituted by the aggregation of basic buildings, but represents an architectural emergency in the city. However, as well known, this theatre had a very short life, it was in fact destroyed just the year following its inauguration, yet the objectives achieved with it were not forgotten; they were rather absorbed by the projects that followed one another.[2] In particular, the project by Felice Giorgi, author of the *Historical Description of the Tor di Nona theatre*, is particularly emphasized; it also retained 6 tiers of 29 boxes each and a total of 678 seats in the stalls; the elliptical curve and the elevations were made with the collaboration with Cosimo Morelli. The theatre inaugurated in 1795 as the Apollo Theatre and for it Valadier towards the Sant'Angelo Bridge erected an architectural facade with three large doors flanked by columns[3] and pillars, crowned by two statues and adorned with the Torlonia coat of arms, all in marble and stucco. The construction of the embankments of the Tiber in 1889 determined its definitive demolition.

Morelli was also the protagonist, together with Antonio Foschini, in the contemporary construction of the Municipal Theatre of Ferrara, inaugurated in 1798 and which has come down to us unlike the Tordinona. The construction site, which lasted over a decade, was started under papal domination and ended at the time of the Cispadan Republic. At the behest of the Pope, the insistent houses on the so-called Isola del Cervo were demolished at night, to carry out a first project by Campana. On the other hand, the execution respects the creation by Morelli, inventor of an interesting correspondence between the two parallel axes of the oval courtyard access for the carriages and of the hall, a covered knotted space, as well as the neighboring oval church of San Carlo built by Aleotti. The building, harmoniously inserted in the consolidated urban fabric, determines a strong interpenetration between the inside and the outside, since the oval courtyard facing along via Giovecca, on which the largest of the two free strong elevations stands with longitudinal accentuation, consisting of a high base in smooth ashlar, windows with a low arched tympanum and an evident cornice. The building also features an interesting corner solution, bevel on Largo Castello, shops along via Giovecca and arcades with the main pedestrian access to the Este castle. The interior is characterized by the absence of the royal box, as well as by the absence of the proscenium boxes, although all four orders of 23 boxes each and the gallery are welded directly to the semi-elliptical arch of the proscenium. A lowered vault covers the room. An interesting debate is linked to the Municipal Theatre of Ferrara regarding the shape that the curve of the theatre should have taken. In particular, Antonio Foschini, whose role in the design has been claimed over time with respect to Morelli, was the author of a *Dissertation on theatrical curves*.[4] The new elliptical curve that Foschini and Morelli proposed to conclude the project already partially realized by Campana, was submitted to the opinions by Piermarini and Stratico. As with the Teatro alla Scala, the Municipal Theatre of Ferrara also appears as an organic complex containing various functions and services, including an entire building dedicated to the artists and communicating with the stage. Over time, various renovations have taken place involving the decorative apparatus, reworked by Migliari, the systems and structures. However, unlike the Scaligero example, the plant has not been heavily altered, and has indeed come almost intact to this day.

In conclusion, we mention the Social Theatre of Mantua, inaugurated in 1822 on a project by Luigi Canonica. This stands on an area at the meeting of seven converging roads from Cremona and Milan, such as to constitute a nodal space at an urban level. The theatre had to be a sort of hinge between the central area of the city, to be enhanced, and the more peripheral rings. The horseshoe shape was adopted, now widespread after the Scala example as an essential character of the type, improved visibility and acoustics (Fig. 5.2). The external façade with a pronaos (Fig. 5.3) resembled the Concordia theatre in Cremona, also performed by Canonica. Several variants were carried out in the executive phase, including the enlargement of the passage under the portico, proposed by the project manager Marconi and shown below (Fig. 5.4), constituting one of the rare drawings preserved in the Mantuan Municipal Historical Archives. The subsequent adjustments and restorations carried out in the 1950s and 2000s by Sodano and Melli have returned, also in this case, as in Ferrara, an unaltered example of Italian theatre.

In summary, we are witnessing a first codification of the external spaces, treated as buildings, examples of specialized construction. The foundation, elevation, unification and conclusion are clearly distinguished; the portico for the transit of carriages

5.1 Public Theatre with Boxes, Italian Theatre

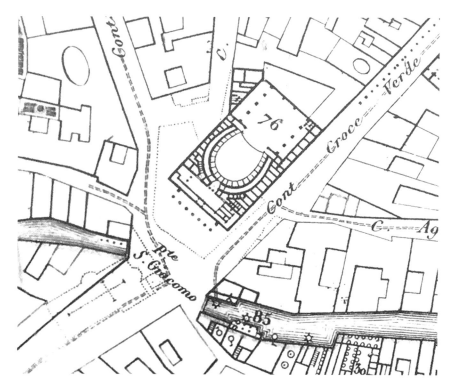

Fig. 5.2 Teatro Sociale in the urban context with demolitions highlighted. The most visible improvements, from the urbanistic point of view, would have been the liberation of the narrow district of San Francesco, the enlargement of the district of Concole, the rectification of the outlet of the district of Pradella on the square, and the disappearance of the alley of Sant'Antonio, which ran from the height of the Agnello district to the Folengo square then occupied by the Oratory of Sant'Antonino, as well as some houses

is established, often realized in the form of a real advanced pronaos, to recall and direct the pedestrian flows at the same time.

5.2 The Synchronic Variant of Court and Academic Theatre

As described so far, the type of Italian theatre was born and consolidated far from the courts and academies. However, the same originally represented a constructive alternative, which obviously took into account the results achieved precisely in the areas from which it was detached. In the following, we analyze some case studies relating to the courts and academies that have continued to influence the type of

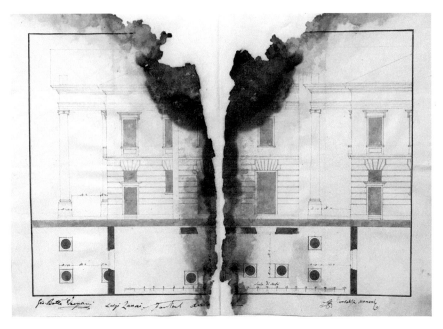

Fig. 5.3 Progetto per l'allargamento del Pronao—Teatro Sociale [Archivio Storico Comunale di Mantova, Titolo XIV—Sicurezza Pubblica, Articolo 24° Teatri, 1824—1831, Suddivisione 2—Teatro Sociale, busta n. 1085]

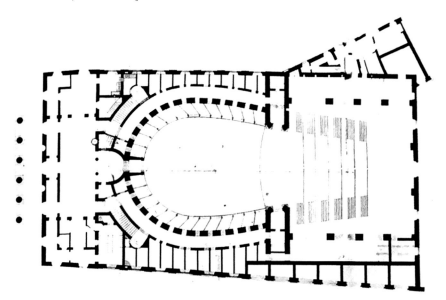

Fig. 5.4 Teatro Sociale—plan [elaboration by the author]

5.2 The Synchronic Variant of Court and Academic Theatre

Italian theatre and that, in some cases, have undergone progressive transformations that have assimilated them, making them in this case, in turn, variants.

The Teatro dei Rinnovati in Siena, reopened in June 2009 after 5 years of restoration conducted by the architect Ettore Vio, was born as an appendix to the Palazzo Pubblico in the fourteenth century. The building in which the General Council Hall was housed, which became a theatre in 1560 after the Medici conquest, featured prisons on the lower floors. The short northwestern front gradually joined the palace starting from the fourteenth century. The main interventions on the wooden room are due to Riccio Senese while a hypothesis of solidification is due to Carlo Fontana.[5] It was destroyed and rebuilt several times due to fires and earthquakes, first by Saverio Posi, author of the Argentina Theatre, and Antonio Donnini, then by Antonio Galli Bibiena. Following the earthquake of 1798, a wall lining over 15 m high was built which covered the facade towards the market square. The layout by Bibiena, bell-shaped with divergent ends connected with the proscenium boxes, has been kept substantially unchanged until today and, also given the public use similar to that of the Italian model, it can be said to be a synchronic variant. Vio's intervention has returned a theatre that has mixed with the municipal offices, new service and connective spaces to improve its usability and functionality, a modern stage and the mystical gulf, as well as having brought to light the masonry arches of conjunction between Palazzo Pubblico and the theatre dating back to the fourteenth century. In its extreme procedural complexity, of which only the essential elements have been reported, the theatre constitutes an extraordinary example of adaptability, a symptom of the continuity of the developing process.

The Teatro della Fortuna di Fano, built by Torelli[6] between 1655 and 1677 to replace the previous hall for comedies, no longer suitable, built by young amateurs inside Palazzo della Ragione about a century earlier, is very close to that of Rinnovati described above. The palace had gradually formed over the span of over six centuries, as a complex comprising a medieval portion, which initially stood isolated, a triple loggia on the ground floor and a large frescoed hall on the upper floor, transformed into a theatrical container. The Theatre of Fano had extraordinary similarities with the Farnese Theatre, whose characters had an immediate and wide diffusion in the Emilia-Veneto area[7] and then towards the other Italian regions. It was in boxes (21 for each of the 5 orders), had a mixtilinear plan in the shape of an apsed rectangle with the long sides slightly diverging towards the proscenium; the back recessed wall was in the shape of a semidecagon; the proscenium, isolated from the boxes, was grandiose and delimited by a giant order of fluted pilasters; finally, it was equipped with a set of scenes created by Torelli and renewed by Bibiena in the early decades of the eighteenth century. In 1839, with the project by the architect Luigi Poletti, a complete makeover of the theatre was carried out, according to the Italian type, which has survived until today. In the seat of a court theatre, there is therefore an updated public theatre, in a horseshoe shape, with a high base to support the boxes, 21 for each of the 3 orders and a gallery. This configuration has been fully preserved despite the numerous interventions carried out up to the present day.[8]

Another example in this sense is the Teatro di Feltre, or della Sena, inside the former Palazzo della Ragione, now called Palazzo del Teatro.[9]

Also of particular importance is the Teatro della Pergola of the Accademia degli Immobili, built in Florence in 1652 by Ferdinando Tacca, who, after visiting the theatres of Ferrara, Mantua, Parma, Bologna and Modena, proposed a variant of the U-shaped plan, so-called ovata, to be built in the disused pulling frame of the wool Art, anticipating some of the main characters of the Italian theatre in the context of a composite academic building. As in the previous examples, the theatre does not achieve full autonomy as it is always contextualized within complex building structures. However, only at the Pergola there is a theatre born already corresponding to the Italian type for the internal configuration aspects of the hall. Initially, in the stalls, men were separated from women, thanks to the steps, arranged at the foot of the first order of boxes and then converted into a further order. This configuration also allowed the execution of exercises with horses. The theatre was opened to the general public in 1718 and this led to its updating: progressively the same boxes were joined to the enlarged stage, the orchestra was lowered, new dressing rooms and distribution spaces were created thanks to the acquisition of two adjoining houses; in particular, Mannaioni[10] oversaw the expansion of the number of boxes from 78 to 84 for a total of 4 orders, as well as the closure of the loggia to create the fifth. The last two orders were then transformed again into galleries in the early twentieth century.

A later example of a theatre born in an academic context and representing a synchronic variant is the Bibiena Scientific Theatre at the Virgilian Academy (Fig. 5.5). In place of the eighteenth-century building in which the hall is contained, there was a building of medieval origin, reworked during the Renaissance, which was Ferrante Gonzaga and his son Cesare's home until 1565. He founded there

Fig. 5.5 Antonio Bibbiena—Teatro Scientifico dell'Accademia [photo by the author]

5.2 The Synchronic Variant of Court and Academic Theatre

the Accademia degli Invaghiti in 1562. At the same time, a first theatre was built consisting of a stage for acting and a wooden staircase. Following the transfer of the Invaghiti to the Ducal Palace, Ferrante II, son of Cesare, invited the Accademia degli Invitti, later known as the Accademia dei Timidi, to his home. In 1766, the last one demolished the sixteenth-century theatre, of which visible traces are preserved in the attic of the current building, to create a larger one, also including adjoining rooms. The new theatre was built between 1767 and 1769 on a project by Antonio Bibiena, for scientific meetings but also for performances and concerts. The stage was given up, replaced by a tribune at the back; the end of the space marked the return to the fixed scene, configured however as a sort of practicable, circular loggia. The effect is of extraordinary organicity and unity. The structure is entirely in brick, there are conspicuous wooden ornamental parts to favor the acoustics; there were no fixed seats in the stalls, just to allow their various use. The external facade is a work by Giuseppe Piermarini, winner of a competition in 1770 announced for the reorganization of the building.[11] The works were carried out between 1772 and 1775, directed by Paolo Pozzo. A further facade was created a posteriori following the demolition of the Church of Santa Maria del Popolo.

The two subsequent examples, at the end of the reflection on the synchronic variants of Italian theatre, constitute a further advancement with respect to what has been described so far. These are the Court Theatre of Caserta and the Carignano Theatre of Turin. These in fact, despite being included in extremely complex and large-scale architectural structures, are configured as nodal spaces and create an autonomous relationship with the natural context and the urban fabric, respectively.

Thus the Court Theatre of Caserta (Fig. 5.6) in the description by Vanvitelli: "*it seems that it gravitates on the columns, with its segments, but in essence it gravitates on the small stages, between one stage and another, which are contrasted by the small vaults of the corridors of the Palchetti, and finally from the exterior of the courtyards, which flank everything, the boxes form lunettes, but these disappear as the vault rises, so that it seems like a crooked shell with a picture in the middle*".[11] The nodal function of the theatrical hall, a space brought and served, is clearly deduced. Inserted in the body of the palace, at the intersection between the central orthogonal arm and the external left side, it contains the structural thrusts in both directions, it is also perfectly integrated into the path system, with accessibility to and from two courtyards. It has a horseshoe shape, which originally continues in the curve of the proscenium producing greater unity. The real floor differs for the presence of the continuous balustrade, compared to the architecturally identified boxes of the upper floors. New elements were represented by the use of the giant order for the entire hall and the stage, equipped with 16 cuts per side and a drum mechanism such as to be able to make scene changes with the curtain open, as well as from the entrance hall, probably used as a foyer, and from the access gallery to the royal box. The opening at the back of the stage allowed the extension of the theatrical event in the garden, similar to what has been described for the Tordinona Theatre. Even this theatre, the only one completed and built by Vanvitelli, is today substantially unaltered since its inauguration in 1769, which was followed by numerous maintenance and conservative restoration interventions, including recent ones.[12]

Fig. 5.6 Luigi Vanvitelli—Teatro di Corte [photo by the author]

5.2 The Synchronic Variant of Court and Academic Theatre

The developing process of the Carignano Theatre began in 1707, with the purchase by Emanuele Filiberto Amedeo di Savoia Carignano of a room for the jeu de paume, in the so-called "trincotto Zaffarone" in Turin. Measuring 12×30 m, it was equipped with windows, wrapped on three sides by pantellere, that is by loggias for spectators, on the fourth side, it had deposits for rackets. His son Vittorio Amedeo di Carignano decided to convert it into a theatre, having modest works carried out between 1709 and 1712 for the construction of a hall with boxes. The entrance took place from the Contrada Nova, while a wall separated the building from the square. In 1749, the facade was in curtain with a small portico composed of two columns near the main entrance. In 1752, Prince Louis entrusted the renovation to Alfieri and bought at the same time some small houses adhering to the border in order to increase the site of the theatre. The Alfieri theatre already has the typical characteristics of the Italian theatre, including a horseshoe curve, extremely modern and diversified compared to the oval truncated designed by Juvarra for the Teatro Regio. The theatre is integrated into the city fabric, the hall is knotted there, as it were, preceded by a C-shaped body that connects it with the urban node represented by piazza Carignano, at the same time creating a correspondence with the overlooking building of the same name, as it can be seen from the *Copy of the map of the interior of the city.* In the aforementioned C-shaped body, some services were integrated on the ground floor, corresponding to the royal apartment on the upper floor. The overall facade of the building in which the hall is included has a width equal to that of Palazzo Carignano; in the middle, not the entrance to the theatre, off-center, stood out, but the princely apartment. The facade has the features commonly used for residential buildings, in line with the example of Sabbioneta. In 1786, the theatre was destroyed by a fire and the reconstruction "as it was and where it was" was entrusted to Giovanni and Francesco Ferroggio. Pietro Carrera at the end of the nineteenth century replaced the fourth tier of boxes with a gallery and created a variation of the portico and entrances, converted into a foyer in the latest restoration by Paolo Morconi. Between the reconstruction of the Ferroggio family and the more recent intervention by Morconi, Bonicelli created two external staircases to facilitate the outflow of the hall, effectively isolating it from the residential building in which it was inserted, he replaced the gallery with a second gallery, modernized the stage, intervened on the systems, services, furnishings and finishes; he finally created a new entrance from via Roma equipped with a reduced to access the stalls, currently intended for artists. Finally, Morconi was responsible for the redesign of the safety stairs and the adaptation of the systems and the scenic tower.

5.3 Main Influences, Inbound and Outbound, Towards the European Context

Undoubtedly, what has been analyzed so far constitutes a point of arrival for most of Italy, however not exempt from European influences and in turn exported, mainly thanks to the work of prominent personalities in the architectural and scenographic fields. Think for example of figures such as Giulio Parigi, Giacomo Torelli, Vigarani and Galli Bibiena. Giulio Parigi's intuitions of the depth of space, of the asymmetry in the scene, of the historical settings as well as of the fantastic ones were characters of great innovation, which spread in Europe as signifiers of the Italian scene, of the baroque scene founded in large part also on the machinery (installed in the backstage, movable bridges and attic). Among the pupils and imitators of Parigi are: Inigo Jones in England, Joseph Furttenbach in Germany, Cosimo Lotti in Madrid, Jacques Callot in France and Stefano della Bella. Giacomo Torelli conceived in 1641 for the Teatro Novissimo, then for the Santi Giovanni e Paolo, the mobile scenes, re-using them in the Venice theatres of San Luca and San Giovanni Crisostomo. *"In France Torelli did not initially have to confront any hypothetical rival capable of obscuring his abilities. The techniques he used since the first show he was called to compete in 1645, La Finta Pazza, were completely new and considered effectively revolutionary in the country where he was welcomed"*.[13] Bibienas are responsible for the corner scenes and the concrete diffusion of Italian theatre throughout Europe and in particular in Austria, France, Spain, Bohemia, Germany, Holland, England, Portugal, Russia and Sweden. Breaking the central perspective focus made the scene wider and the actors were no longer relegated to the proscenium alone, becoming, on the contrary, part of the scene itself, at least for considerable portions.

The Shakespearean theatre, an alternative model to the Italian theatre, was therefore overwhelmed by the latter during the seventeenth century. Also through the work by Inigo Jones, England converted to Palladio, the wings prevailed over the periactis, the galleries were replaced by boxes. Further on, the Berlin Opera House, in 1742, was built halfway between the court theatre and the public theatre, not even Boullée seems to completely renounce the boxes in his ideal design for the opera house.

Between 1872 and 1876, the Bayreuth Festspielhaus was completed according to Wagnerian requirements: the orchestra is located in the mystical gulf, the seats recall the layout of an amphitheatre, boxes and galleries are absent, the public remains in the shade. Many have recognized in this theatre the end of the Italian type, the birth of a new model; yet, on closer inspection, the clear distinction between the hall and the stage is taken to its extreme consequences, realizing the Renaissance ideal of the perspective scene with the utmost rigor. The Bayreuth Theatre by Semper is therefore not an overturning, but a vertex of the theatrical tradition. Milizia had recognized the perspective scene as an antivitruvian aspect par excellence, as the unification of the scenic framework created nothing but an illusion in the viewer. The spatial unity between him and the actor was irretrievably lost, a desire that reappears later with the Gropius model. The total theatre, as we know, was never built and failed to produce a unanimous movement.[14] The same proposed three alternative configurations of which the first classic and the other two consisting, respectively, of the central plan, typical of arenas and with the central stage, the so-called thrust developed in the contemporary world by Izenour. The substantial abolition of the boxes to improve the usability of the shows, a phenomenon recorded almost everywhere in recent interventions, alone is not sufficient to mark a clear break with the past, since contemporary dramatic theatre remains faithful to the traditional perspective scene. Currently, as mentioned at the beginning, there are new theatrical programs but not adequately supported by equally new spaces for the show. Further spaces are sought, returning to the city, and in some cases even contesting the non-traditional places themselves. Without abandoning the privileged field of research, the evolutions of the characters of modern theatre in the Italian context will be discussed in the following in-depth studies, in which, as seen so far, the public theatre with boxes, in the Italian type, has known an unusual longevity, being in fact, the largest number of historical theatres in Europe and in the West in general are still active, without any prejudice.

5.4 The Specialization of Building Types

To describe the return to the city, the grafting into the typical urban fabric again and above all in contemporary theatre, we analyze cases belonging to the Milanese context, such as Palazzina Liberty, the San Babila Theatre and Piccolo.

Palazzina Liberty (Fig. 5.7), designed in 1908 by the architect Migliorini, the central building of the old fruit and vegetable market from 1911 to 1965, was built on an area owned by the municipality. It contained trading rooms, a restaurant and even a doctor's office. It was occupied and adapted to a theatrical space in 1973 by Dario Fo and Franca Rame. The works carried out in a good way produced a hall with 800 seats. Exactly a decade later Dario Fo returned the keys of the building to the Municipality

Fig. 5.7 Palazzina Liberty, Milano

and the room ended up in decay again. The following year the renovation project was approved by the architects Piero De Amicis and Paolo Giussani, and by the engineer Giuseppe Introzzi. Completed the works in 1992, today the building is the seat of the Civic Wind Orchestra, hosts concerts and cultural events. It looks like a bi-apsidal rectangle, with large glass surfaces and elegant decorations. The main compartment, functionally served and structurally brought, was adapted to a hall, containing the stalls and the stage, in the lateral distribution areas some galleries were obtained, as can be seen from the images shown. Being an example of specialized construction, the procedural transformation took place in absolute continuity.

Between 1890 and 1892 Palazzo Carmagnola, one of the oldest buildings in the city, underwent a radical restoration, in accordance with the changes envisaged by the Master Plan for via Dante. The main courtyard lost one side. New restorations, scarcely attentive to the glorious past and to the architectural value of the building, were carried out between 1938 and 1939, to adapt the structure to the Civic Dopolavoro. Inside the building, served by the minor courtyard, a space was created for performances. Today Piccolo Teatro Grassi has kept its modest entrance on via Rovello, 2 unchanged since 1940. It was inaugurated again with the Scala orchestra in 1947 and has a small hall of just 650 seats, renovated in 1952 based on a project by architects Rogers and Zanuso. The stage was enlarged, a parallel gallery was created, the ceiling was shaped according to specific measures aimed at improving the acoustics, the furniture was updated with design objects, including the classic Zanuso self-referential armchair; the Municipality also granted five additional premises to be used as offices. Thus, the Piccolo Teatro in Milano was born on the former Dopolavoro Civico, which became a third viewing cinema, bivouac during the Republic of Salò and a place of entertainment for soldiers of the allied army. The last restorations of 2008 have brought to light conspicuous testimonies of the glorious past, including the confirmation of the presence of a large room used for carnival performances, and have allowed the new and complete use of the minor courtyard, used as a market for wheat and flour, then as a garage and parking area until that date. Today along the courtyard there are small exhibition rooms, a bookshop and a bar (Fig. 5.8).

Fig. 5.8 Marco Zanuso—Il Piccolo Teatro Grassi and Il Piccolo Teatro Strehler in Milano [photo by the author]

5.4 The Specialization of Building Types

Fig. 5.9 Teatro San Babila [photo by the author]

Despite all the interventions mentioned, also relating to the previous case, have taken place quite recently, we can identify a process that is very similar to that described for theatres born within complex building structures.

The San Babila Theatre (Fig. 5.9), built in 1964 according to a design by Maria Gottardi in collaboration with Luigi Danova Belisario Duca, is another example of this. It has 474 seats of which 342 in the stalls and 132 in the gallery, the plan is that of an irregular mixtilinear polygon that gradually decreases towards the top. Also in this case, the public theatre system is therefore preserved exclusively as regards the establishment of the traditional relationship between stage and actual hall.

The Teatro Studio (Fig. 5.10) constitutes, with respect to the aforementioned cases, an advancement as it can be considered not a simple appropriation of the pre-existence to innovate it with the introduction of a known type but rather the evolution of the type itself. It represents the adaptation of the pre-existing Fossati Theatre by Fermo Zuccari, dating back to 1859, originally conceived as a daytime theatre, without a roof and equipped with two baroque facades designed by the Boni brothers. Inserted in an urban block and therefore free only on two fronts, it presented an Italian type layout. Zanuso takes up the planimetric conformation of the horseshoe curve only, along which he creates steps and galleries of previous ancestry compared

Fig. 5.10 Teatro Studio, Milano [photo by the author]

to the Italian type. The stage is on the same level as the so-called theatre square. An open and multipurpose space is thus created, which creates the thrust. Inaugurated in 1986, it has a total capacity of approximately 500 seats for an area of 33×19 m. The European Higher School of Theatre is also located inside it. The building is connected to the Teatro Grande by an underpass useful for accessing common services.

Finally, mention is made of the last headquarters of the Piccolo (Fig. 5.11), built in a nerve center for the city of Milan, not far from the Brera Museum, the Botanical Garden and the Sforza Castle, designed by Marco Zanuso with the help of the set designer Ezio Frigerio. The capacity of the theatre is 1200 seats, it is equipped with services, which have led to the definitions of "theatre factory", polymorphic complex, "a new theatre for a new theatre". The construction works were undertaken between 1979 and 1980 and completed in 1998; the plant is an irregular octagon, the core of which consists of a large stage of about 17×23 m, there is the orchestral pit. The intent was to create a new theatre with adequate audiences and means but which would retain the imprint of the theatre in via Rovello, which would represent the rite of the show but also the rite of behind-the-scenes production work. The system can be seen as the geometric intersection of two squares each relating to vision, consumption of the show and work, respectively. The building, in which the scenic tower partially emerges, made up of externally closed volumes, in curtain, is knotted around the theatre hall. The nodality is also transferred to the urban level. On the basis of the developing process, the new building creates an autonomous and nodal specialist building, fully expressive of the characteristics of the type.

5.4 The Specialization of Building Types

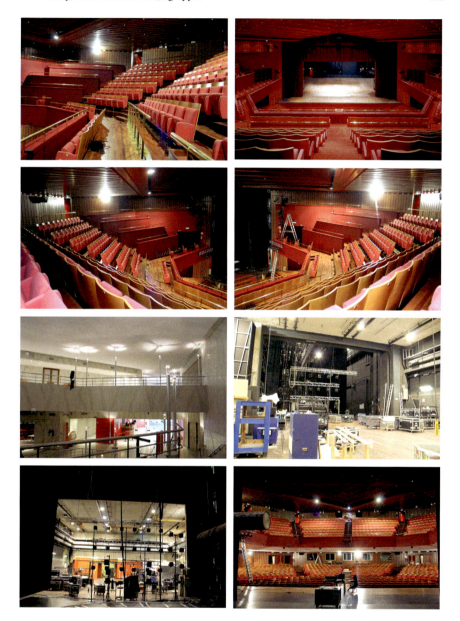

Fig. 5.11 Marco Zanuso—Il Piccolo di Milano [photo by the author]

5.5 Summary

We can say that we have investigated the development of the characters of the Italian theatre by analyzing some significant case studies among the national projects up to recent times.

The public theatres with boxes distinguished themselves in particular by the synchronic variants constituted by the court and academic theatres.

The main incoming and outgoing influences of the type at a European level were then mentioned.

In closing, thanks to the analysis of recent projects carried out in the city of Milan, it was emphasized that the process of grafting into the urban fabric of the public theatre with the consequent specialization of construction is in fact still active.

5.5 Summary

TEATRO TOR DI NONA — CODE5.01

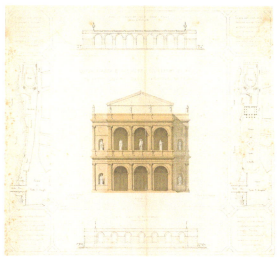

Nuova Piazza e Prospetto del Teatro Apollo con rettificazioni dell'attuale Vestibolo [Archivio di Stato di Roma, Collezione I di disegni e mappe, cartella 89, foglio 631]

TYPOLOGICAL SCHEME

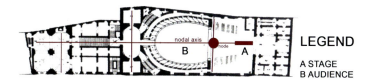

LEGEND
A STAGE
B AUDIENCE

SUMMARY OF THE MAIN SELECTION CHARACTERS

place_**ROMA**
level of the event_**national**
site of di architectural interest_**Yes**
UNESCO site_**Yes**
year_**1671**
building_**permanent theatre**
construction_**wooden and brick**
features_**proscenium**
keyword_**architectural node**

TEATRO COMUNALE DI FERRARA — CODE5.02

Foschini, Morelli - Teatro Comunale di Ferrara - 2017

TYPOLOGICAL SCHEME

LEGEND

A STAGE
B AUDIENCE

SUMMARY OF THE MAIN SELECTION CHARACTERS

place_**FERRARA**
level of the event_**regional**
site of architectural interest_**Yes**
UNESCO site_**Yes**
year_**1797**
building_**permanent theatre**
construction_**mixed structure**
features_**proscenium**
keyword_**architectural node**

5.5 Summary

TEATRO DELLA PERGOLA — CODE5.03

Ferdinando Tacca - Teatro della Pergola - 2017

TYPOLOGICAL SCHEME

LEGEND

A STAGE
B AUDIENCE

SUMMARY OF THE MAIN SELECTION CHARACTERS

palce_**FIRENZE**
level of the event_**national**
site of architectural interest_**Yes**
UNESCO site_**Yes**
year_**1652**
building_**permanent theatre**
construction_**mixed structure**
features_**proscenium**
keyword_**knotting**

TEATRO DI CORTE — CODE5.04

Luigi Vanvitelli - Index and cut of the middle part of the Royal Palace in its greater extension (detail) - 1756

TYPOLOGICAL SCHEME

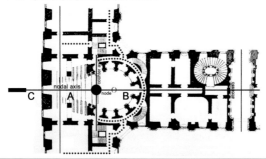

LEGEND

A STAGE
B AUDIENCE
C COURTYARD
(natural scene)

SUMMARY OF THE MAIN SELECTION CHARACTERS

place_**CASERTA**
level of the event_**national**
site of architectural interest_**Yes**
UNESCO site_**Yes**
year_**1769**
building_**permanent theatre**
construction_**mixed structure**
features_**proscenium**
keyword_**architetctural node**

5.5 Summary

TEATRO CARIGNANO — CODE5.05

Benedetto Alfieri - Teatro Carignano - 2017

TYPOLOGICAL SCHEME

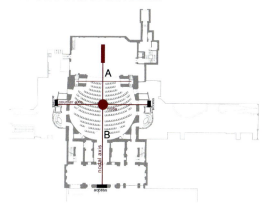

LEGEND

A STAGE
B AUDIENCE

SUMMARY OF THE MAIN SELECTION CHARACTERS

place_**TORINO**
level of the event_**national**
site of architectural interest_**Yes**
UNESCO site_**No**
year_**1753**
building_**permanent tehatre**
construction_**mixed structure**
features_**proscenium**
keyword_**architectural node**

PICCOLO TEATRO GRASSI — CODE5.06

Rogers, Zanuso - Entrance of Piccolo Teatro Grassi - 2017

TYPOLOGICAL SCHEME

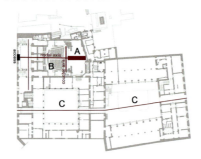

LEGEND

A STAGE
B AUDIENCE
C COURTYARD

SUMMARY OF THE MAIN SELECTION CHARACTERS

place_**MILANO**
level of the event_**national**
site of architectural interest_**Sì**
UNESCO site_**No**
year_**1947**
building_**permanent theatre**
construction_**mixed structure**
features_**proscenium**
keyword_**knotting**

5.5 Summary

PICCOLO TEATRO STUDIO — CODE5.07

Marco Zanuso - Piccolo Teatro Studio - 2017

TYPOLOGICAL SCHEME

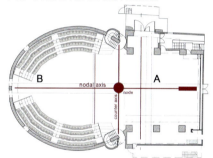

LEGEND

A STAGE
B AUDIENCE

SUMMARY OF THE MAIN SELECTION CHARACTERS

palce_**MILANO**
level of the event_**national**
site of architectural interest_**Yes**
UNESCO site_**No**
year_**1986**
building_**permanent theatre**
construction_**mixed structure**
features_**thrust**
keyword_**knotting**

PICCOLO TEATRO STREHLER — CODE5.08

Marco Zanuso - Piccolo Teatro Strehler - 2017

TYPOLOGICAL SCHEME

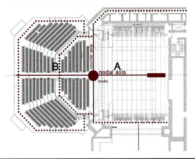

LEGEND

A STAGE
B AUDIENCE

SINTESI DEI PRINCIPALI CARATTERI DI SELEZIONE

place_**MILANO**
level of the event_**national**
site of architectural interest_**Yes**
UNESCO site_**No**
year_**1999**
building_**permanent theatre**
construction_**reinforced concrete**
features_**proscenium**
keyword_**architectural node**

5.5 Summary

TEATRO TOR DI NONA — CODE5.21

Category	Option	first design by FONTANA	second design by FONTANA	design by GIORGI
stage	flat	□	□	□
	inclined	■	■	■
scene	fixed	□	□	□
	mobile	■	■	■
	material	□	□	□
	perspective	■	■	■
scaenae frons	present	□	□	□
	absent	■	■	■
proscenium	present	■	■	■
	absent	□	□	□
separation audience/stage	present	□	□	□
	absent	■	■	■
audience	"U" shape	■	□	□
	"V" shape	□	□	□
	jagged	□	□	□
	bell shape	□	□	□
	circular	□	□	□
	oval	□	□	□
	elliptical	□	□	□
	double ellipse	□	□	□
	horseshoe	□	■	■
	mirrored	□	□	□
	flat	□	□	□
	inclined	■	■	■
orchestra	bounded	■	■	■
	not bounded	□	□	□
	equal	■	■	■
	lower	□	□	□
theatre square	present	■	■	■
	absent	■	■	■
seats	with steps	□	□	□
	with steps and loggias	□	□	□
	with steps and boxes	□	□	□
	with boxes	■	■	■

COMPARATIVE TABLE — CODE 5.29

SINGLE SCENE DISTINCT FROM THE AREA DEDICATED TO SPECTATORS
ARCHITECTURAL NODE
CENTRAL, CURVE WELDED TO THE STAGE, PROSCENIUM, DOUBLE AXIALITY

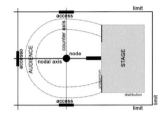

SYNCHRONIC VARIANTS

ARCHITECTURAL NODE
NOT SYMMETRICAL, AT THE PROSCENIUM AND THE ORCHESTRA, CURVE NOT WELDED TO THE STAGE, DOUBLE AXIALITY

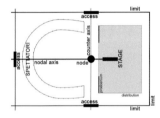

ARCHITECTURAL NODE
THRUST

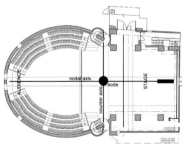

Notes

1. Described by Pevsner N (1976) A history of building types. Princeton University Press, Princeton.
2. Following the demolition ordered by Innocenzo XII in 1698, the theatre was rebuilt in 1733 on a project by Gregorini and Passalacqua by order of the new Pope Clemente XII at the expense of the Apostolic Chamber. This realization included an almost circular hall with four tiers of boxes. In 1764,

5.5 Summary

it was reopened after a fire, with six tiers of boxes and respectful of the previous works. Adjustments were also made in 1768 for the construction of ten proscenium boxes and in 1769 for the construction of a spur on the Tiber. This theatre too was destroyed by a fire in 1780. In 1782, Tarquini's project was partially realized, then demolished because it was structurally unstable; Barberi's project followed.

3. The two columns of caristio marble belonged to the recently discovered Villa dei Quintili on the via Appia.

4. ms. in FEc CI. I, n. 618.

5. It had a proscenium framed by a giant order of Corinthian pilasters and a fixed scene; the hall was U-shaped, with 27 boxes for each of the 4 orders, divided by curved partitions, there were then benches nailed to the floor in the stalls for a total capacity assumed of 1535 seats; the last order was made up of open boxes interspersed with shelves.

6. In 1665, the petition to the General Council presented by 17 Fano patricians including Torelli to obtain the concession for the construction of the theatre.

7. Think of Teatro della Sala and Teatro Formagliari in Bologna.

8. At the end of the nineteenth century, the central boxes of the third order were transformed into a gallery by breaking down the dividing walls, to increase their capacity and to "displace the stalls", without however altering the typological conception of the theatre. Following a violent earthquake in 1930, work was also carried out on the construction of the pit for the orchestra, which reduced the space of the stage, which was already shallow. This aspect was obviated a few years later with the demolition and reconstruction of the jetty connecting with the Malatesta Palace designed by Torelli. The atrium was also enlarged, occupying two of the three innermost arcades initially characterizing the building. In 1944, the theatre was demolished by the collapse of the neighboring civic tower, the consequent restorations carried out after the war led to the total loss of the painted vaulted roof but preserved the layout of the theatre.

9. This, built between 1549 and 1583, housed the hall for the meetings of the Maggior Consiglio which, in turn completed in 1610, was never inaugurated because it was too large to heat and unsuitable for its function. Instead, it was used as a hall for parties and for theatrical performances, especially in carnival time. In 1684, the Social Theatre with boxes was built; in 1741, a third tier of boxes chained to the ceiling was added and the two lower tiers were consolidated. Selva visited Feltre to propose a makeover of the theatre in 1802, with a reduction in the number of boxes for optical reasons. Finally, the works started in 1811 with an increase in the number of boxes, the Social Theatre was maintained until 1899, then municipal. Closed in 1929 because it no longer complied with safety standards, it was the subject of a recently completed restoration campaign supported by Italia Nostra.

10. Antonio Bibiena's project was rejected because it was deemed too ornate, Bibiena will therefore propose it again for the Bologna Theatre. During the nineteenth century, various adaptation works were also carried out, including

126   5 Theatre as Urban Knot

the construction of a hall with a gallery, new dressing rooms, the lowering of the stalls and stage, new service rooms.

11. Both the old Bibiena and Crevola were defeated.
12. Jacobitti G M (1995) Un gioiello ritrovato. In: Il teatro di corte di Caserta. Electa, Napoli: 7.
13. The decorative apparatus was completed between 1768 and 1769. In 1805, following an earthquake and, in 1812, following a tornado, by Giuseppe Cammarano, the royal box was restored. The theatre was active again between 1816 and 1824 with the restoration; in 1850, it was already in a bad state of conservation. It suffered further damage with the 1930 earthquake, reopened after the war at the end of a first restoration campaign, followed by a second one between 1989 and 1994.
14. See Milesi F (2000) Giacomo Torelli. L'innovazione scenica nell'Europa barocca. Fondazione Cassa di Risparmio di Fano, Fano.

References

(1937–1946) Il teatro di Feltre. Archivio storico di Belluno, Feltre e Cadore
(1995) Il teatro di corte di Caserta. Electa, Napoli
(2000) Lo "spettacolo maraviglioso". Il Teatro della Pergola: l'opera a Firenze. Edizioni Polistampa Firenze, Firenze
Alberti M, Bartolini A, Marcelli I (2010) L'Accademia degli Immobili. "Proprietari del Teatro di via ella Pergola in Firenze. Inventario". Ministero per i Beni e le Attività Culturali. Direzione Generale per gli Archivi, Roma
Amadei G (1973) I centocinquant'anni del Sociale nella storia dei teatri di Mantova. CITEM, Mantova
Battistelli F (1968) Il teatro della fortuna Appunti per Una Monografia. Fano (4)–169 214
Battistelli F (1972) L'antico e il nuovo teatro della fortuna di Fano (1677–1944). Sangallo, Fano
Battistelli F (1986) Piante delle scene di Ferdinando Bibiena per l'antico Teatro della Fortuna e "annotazioni per chi opera" in un manoscritto inedito del secolo XVIII. Nuovi studi fanesi (1):131–152
Battistelli F (1987) Scritti e carteggio sul sipario di Francesco Grandi per il teatro della Fortuna. Nuovi studi fanesi (2):165–220
Bazzotti U (2007) Il Teatro Scientifico a Mantova. Skira, Milano
Bazzotti U (2015) Il Teatro Scientifico dell'Accademia di Mantova. Il Rio, Mantova
Bragaglia L (1981) Teatro della Fortuna de Fano. Vita italiana (26):123–134
Cametti A (1939) Il teatro Tordinona poi Apollo. Arti Grafiche Aldo Chicca, Tivoli
Canella G (1966) Il sistema teatrale a Milano. edizioni Dedalo, Bari
Canella G (ed) (1989) Zodiac: Theatre History and Design (2)
Cirani P (2001) Musica e spettacolo nel Teatro Nuovo di Mantova. Edizioni Postumia, Mantova
Garms J. (2014) Reggia di Caserta. Teatro di Corte. Studi sul Settecento Romano (30):169–173
Giovagnoli A (1987) Disegni per il teatro della Fortuna di Fano alla Biblioteca "Luigi Poletti" di Modena. Nuovi studi fanesi (2):111–164
Moreschi A (2013) Il teatro Sociale di Mantova. Fondazione, cronologia, vissuto. editoriale sometti, Mantova
Palmucci Quaglino L (2012) Dal "trincotto Zaffarone" al teatro per il principe di Carignano. In: Cornaglia P, Kieven E, Roggero C (ed) Benedetto Alfieri. Campisano Editore, Roma

References 127

Penna A (1998) Il primo teatro pubblico di Roma: le vicende del teatro Tordinona nel XVII secolo. Studi romani (46):337–368

Piccini G (1912) Il Teatro della Pergola. R. Bemporad & figlio, Firenze

Rebaudengo A (2009) Teatro Carignano. EBS, Verona

Soldini J (1984) Per la storia del neoclassicismo a Mantova: la costruzione del teatro sociale dell'architetto Luigi Canonica attraverso i documenti dell'archivio storico. Bollettino d'Arte del Ministero per i Beni Culturali e Ambientali (23):79–88

Strappa G (2019) The "knotting" as a morphological phenomenon. An interpretation of the Italian Chamber of Deputies forming process. ICONARP International Journal of Architecture & Planning (Special Issue – december 2019)

Strappa G (1994) Tipologia degli organismi seriali e nodali nel ciclo progettuale dei palazzi di giustizia di Gianfranco Caniggia. Bollettino della Biblioteca del D.A.A.C. (maggio)

Tamburini L (2002) Il Teatro D'Angennes di Torino a metà Ottocento. Studi piemontesi XXXI:41–52

Vigni L, Vio E (2010) Storia e restauri del Teatro dei Rinnovati di Siena. Dal consiglio della Campana al salone delle commedie. Pacini Editore, Pisa

Zangheri L (1970) Ferdinando Tacca e il primo teatro all'italiana. Necropoli (11–12):61–69

Documents

Teatro Alibert, detto "delle Dame" (1731–1838) [Archivio di Stato di Roma, Camerale III (secoli XV–XIX), busta n. 2126]

Teatro Apollo, già Tordinona (1669–1861) [Archivio di Stato di Roma, Camerale III (secoli XV–XIX), buste n. 2127, 2128, 2129, 2130]

Teatro Comunale - Brano di relazione intorno all'operato dell'architetto Campana. Attivo e passivo del Teatro Comunale. Nota dei proprietari dei palchi nel dic. 1802. Calcoli intorno alla curva teatrale e note relative alla fabbrica suddetta—sec. XIX [Biblioteca Comunale Ariostea, Fondo Antolini, busta n. 69]

Piante due del Teatro Comunale di Ferrara—sec. XIX [Biblioteca Comunale Ariostea, Fondo Antolini, busta n. 70]

Foschini Antonio—Biglietto con cui dichiara al march. Camillo Bevilacqua di eseguire la curva del Teatro come allo schizzetto inviatogli, 15 agosto 1791. Lettera al March. Bevilacqua intorno al consegnare la pianta del Teatro ai costruttori imolesi, 11 luglio 1791. Medaglioni pubb. nel 1816, incisore F. Padovani—sec. XVIII [Biblioteca Comunale Ariostea, Fondo Antolini, busta n. 73]

Copia di una lettera del principe Gabrielli a d. Rodolfo Varano intorno alla curva del Teatro di Tordinona fatto dal Morelli, 13 giugno 1791—sec. XVIII [Biblioteca Comunale Ariostea, Fondo Antolini, busta n. 74]

Pareri degli ing. e prof. Piermarini e Stratico sulla curva del Teatro Comunale di Ferrara e la distribuzione dei palchi—sec. XVIII [Biblioteca Comunale Ariostea, Fondo Antolini, busta n. 76]

Schizzetti tre della curva del Teatro Com. di Ferrara. I. Distribuzione de' palchi del 1° ordine che si trovava in questo Teatro vecchio. II. Distribuzione dei palchi del 1° ordine secondo la nuova riforma; conservata la solita distanza dalla scena a quelli che ne hanno la proprietà giusta al primo sistema. III. Distribuzione de' palchi del 1° ordine in questo Teatro secondo la nuova riforma conservata la solita distanza dalla scena—sec. XVIII [Biblioteca Comunale Ariostea, Fondo Antolini, busta n. 77]

Saggio sopra l'architettura teatrale, ossia dell'ordinanza la più vantaggiosa ad una sala di teatro relativamente ai principii dell'ottica e dell'acustica—sec. XIX [Biblioteca Comunale Ariostea, Fondo Antolini, busta n. 84]

Teatro Comunale di Ferrara—Circolari, progetti, contestazioni, imprese—sec. XIX [Biblioteca Comunale Ariostea, Fondo Antolini, busta n. 85]

Teatro Comunale di Ferrara. Lavori 1823–1851 [Archivio Storico Comunale di Ferrara, Commissione Pubblici Spettacoli—Direzione Teatrale, busta n. 89]

Teatro Comunale di Ferrara restaurato l'anno 1851 dal celebre artista Prof. Francesco Migliari e successiva morte di questi—sec. XIX [Biblioteca Comunale Ariostea, Fondo Antolini, busta n. 120]

Teatro Comunale di Ferrara. Lavori 1852–1885 [Archivio Storico Comunale di Ferrara, Commissione Pubblici Spettacoli—Direzione Teatrale, busta n. 90]

Teatro Comunale di Ferrara [Archivio Storico Comunale di Ferrara, Mappe, stampe e disegni, 1/10 dal 12 al 61 edifici di interesse storico]

Mappe del Catasto Teresiano [Archivio di Stato di Mantova, Catasto Teresiano, 1771–1888]

Mappe del Catasto Lombardo-Veneto [Archivio di Stato di Mantova, Catasto Lombardo-Veneto, seconda metà del secolo XIX]

Suddivisione 2—Teatro Sociale [Archivio Storico Comunale di Mantova, Titolo XIV—Sicurezza Pubblica, Articolo 24° Teatri, 1824–1831, busta n. 1085]

Suddivisione 4—Teatro Scientifico, Fascicolo n.1 [Archivio Storico Comunale di Mantova, Titolo XIV—Sicurezza Pubblica, Articolo 24° Teatri, 1823–1862, busta n. 1087]

Teatro Accademico [Archivio Storico Comunale di Mantova, Categoria XIV—Sicurezza Pubblica, Classe 4 teatri, spettacoli e Divertimenti Pubblici, Articolo 1 Teatri Concerti, Rappresentazioni e Produzioni Artistiche, 1909–1930]

Chapter 6
The Evolution of Theatre Features

Urban and Architectural Knots

Abstract This chapter investigates the development of the characters of the Italian theatre up to the achievement of urban centrality, up to the most recent redevelopment, recovery and restoration projects, especially in the historical fabric, carried out on the national territory or left on paper, first of all that by Carlo Scarpa for the Carlo Felice of Genoa. We will try to identify the recurring aspects both on the architectural scale and on the urban scale, answering two main questions: referring to the Italian hall, today can we exclusively speak of transformation of the existing? Aluenti, Botta, Rossi tried to modify the figurative language of theatre from inside without denying it, without abandoning the known space, returning to the codified type very close to the Renaissance use of the town square or the courtyard of the stately building. Tafuri recognizes in particular in Aldo Rossi's Carlo Felice some of the characters anticipated by Palladio in the Olympic Theatre, and by Scarpa himself, first of all the recreating an architectural space, expressive of the urban dimension, and not purely illusive beyond the fourth wall. And again: what are the main reasons for the success of the typology, such as to decree its persistence even in contrast with the European and Western context in general?

Keywords Carlo Scarpa · Aldo Rossi · Urban knots · Architectural knots

6.1 A Theatre as an Urban Knot Grafted in the Historical Fabric

The project for the Carlo Felice Theatre in Genoa, severely damaged during the last stages of the Second World War, kept Carlo Scarpa busy for several years at the end of his life. However, the project never abandoned paper. It contains and brilliantly resolves some key issues in the approach to the object of grafting into the historical fabric. Indeed the theatre is designed to become a real node between the ancient city, the nineteenth-century city and the more recent expansion of the center. By overturning the traditional Italian theatre hall in the foyer, a very special and distinctive new public space is obtained: theatre stages theatre, introduces itself to the city. The study aims to retrace the two main phases of Carlo Scarpa's project, highlighting which methodological aspects are still valid today in the process of

© Springer Nature Switzerland AG 2022
S. Clemente, *Reconstructing Theatre Architecture*, The Urban Book Series,
https://doi.org/10.1007/978-3-030-89968-4_6

integration into the historical fabric, and which have marked an evolution in the developing process of the modern space for the show.

Luka Skansi provides a careful examination of the state of affairs of Carlo Felice, with which Carlo Scarpa had to compete in the conception of a general solution, then reworked after about 6 years, and finally changed again.

The war destructions made the structures of the Barabinian theatre almost unusable. The survey of the state of affairs carried out in 1951 shows how the two monumental facades (the pronaos—the ancient royal entrance—towards piazza De Ferrari and the front towards via *XXV aprile), the wall structure of the theatrical hall (which develops with the traditional bell plan) and some internal spaces, in particular the vestibule of the ancient entrance, with the eight isolated columns that give rhythm to the tripartite space. [...] resisted although severely damaged. On the other hand, the scenic tower and the ancient foyer located at the first floor above the entrance from* via *XXV Aprile are unusable, as most of the vertical connections that led spectators to the upper levels of the boxes. [...] The theatre has its own building front closed by the monumental buildings of piazza De Ferrari with its longitudinal front. Indeed, Barabino places on this front the royal entrance with the pronaos characterized by six fluted Doric columns [(Figs. 6.1 and 6.2)]. However, this view does not correspond to the axis of the theatre, which instead finds its natural front on* via *XXV aprile. [...] With the Town Plan of 1954, new urban expectations are added to the theatrical building. The Carlo Felice is, in practice, in a condition of junction*

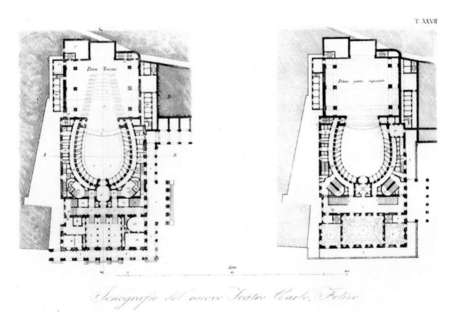

Fig. 6.1 Carlo Barabino-Internal plant of the theatre [https://architettura.unige.it/e-books/berlendis/berlen_3.htm]

6.1 A Theatre as an Urban Knot Grafted in the Historical Fabric

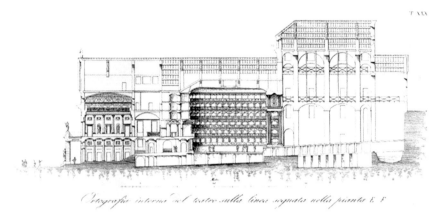

Fig. 6.2 Carlo Barabino—Hymnographies of the new Teatro Carlo Felice [https://architettura.unige.it/e-books/berlendis/berlen_3.htm]

between the ancient city, the nineteenth-century city (between the great axis of viale XX Settembre and Mazzini Gallery) and the new extension of the historic center.[1]

Although Scarpa's project remained on paper, you would wrongly speak of "failure", since some of his most brilliant intuitions were inherited in the later realized work by Aldo Rossi.

So, the architect about the Carlo Felice: *"There is a very interesting project by Carlo Scarpa—presented long before the competition in which I participated—which completely kept the pronaos and then developed into the actual theatre and its side parts"*.[2] Note that, on the contrary, the destruction of the Barabino's pronaos was common to all the projects presented in the competition at the time. And again: *"The problem I posed was that of trying to embody, to assume this historical reality, in a new reality, in a new representation that would automatically draw strength from the existing, almost archaeological elements. [...] The main innovation for what concerns the city was to create, in the area of the ancient entrances, a continuity between Piazza De Ferrari and Mazzini Gallery—now restored and full of shops— which is on the other side. This passageway is a sort of covered square partly illuminated by a zenith light that comes from above through a conical element that crosses and also illuminates the various foyers. One of the characteristics of the project is the overcoming of the classic eighteenth-century plan with boxes, which perhaps for many reasons no longer works and which will instead be re-proposed in the reconstruction project of La Fenice Theatre. This is because the stage is an institution of the nineteenth-century aristocracy which considerably falsifies the calculations relating to the number of spectators because, if in the past the owners shared them with bodyguards or young children, today they are occupied only by two or three spectators"*.[3]

The Carlo Felice Theatre in Genoa therefore never came to completion according to Scarpa's project, like no other theatre by the architect. Only a few limited demolition works for the Genoa theatre marked the start of a construction site that was soon interrupted.

The practice of continually rethinking works up to the construction phase, and actually even while construction was underway, was, moreover, almost a distinctive feature of the way in which Scarpa understood the practice of the architect. This is why the existing works can undoubtedly be considered something quite far from his real ideas and intuitions. Especially since theatre was a completely new theme compared to the architect's professional experience.

Over 700 sheets describe three successive stages of the project development, of which the presentation of the first version would coincide with the inauguration of Castelvecchio. Behind every perspective view there are therefore hundreds of technical drawings, perhaps less fascinating but resolving concrete problems (Figs. 6.3, 6.4 and 6.5).

In particular, from the Archivio dell'Ufficio Speciale presso gli uffici tecnici del Comune di Genova, there are: a first version produced for the building permit of October 5, 1964, a second version dating back to June 3, 1970, and a third version (variant of the second) of June 29, 1973. The aforementioned variants will mainly

Fig. 6.3 Carlo Scarpa—schizzo di studio [Centro Archivi di Architettura—MAXXI, Fondo Carlo Scarpa, Attività Professionale, Progetti e incarichi professionali, 214: Progetto Teatro Carlo Felice, Genova (1963–1981), n.18576]

6.1 A Theatre as an Urban Knot Grafted in the Historical Fabric

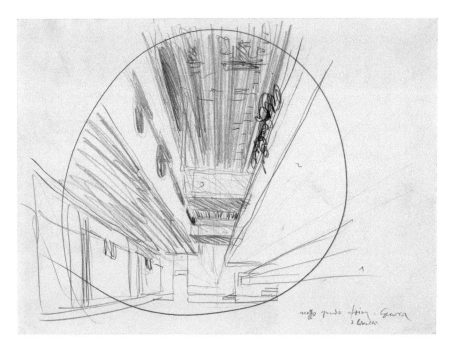

Fig. 6.4 Carlo Scarpa—studio del foyer [Centro Archivi di Architettura—MAXXI, Fondo Carlo Scarpa, Attività Professionale, Progetti e incarichi professionali, 214: Progetto Teatro Carlo Felice, Genova (1963–1981), n.18673R]

affect the foyer, its dimensions, its orientation and its entrances, as well as the conception of the two main fronts, one towards Mazzini Gallery, and the other towards piazza Piccapietra. The theatre could indeed be understood as a real filter between the two urban spaces of the gallery and the square.

The pronaos, in addition to serving as an entrance to the building from the square, becomes the backbone of an urban pedestrian path that crosses the theatre and leads to the portal of the nineteenth-century gallery on piazzetta Labò. […] At the center of the transversal path of the building, still belonging to the domain of the pedestrian flow of the city, the entrance to the reduced is placed, which is thus completely separated from the rest of the functional machinery of the theatre. This seems to be the main reason why the two monumental entrances to the theatre are transformed into junctions: on the one hand they both serve the atrium and the subsequent internal distribution system, on the other they connect to the underlying extension of the gallery and to the small hall. […] Before entering the hall, the spectator has the opportunity to look out over Mazzini Gallery for the last time. At the end of the atrium there is a curious doubled ramp that initially leads to a landing that protrudes from the line of the facade. Here the visitor has the opportunity to glimpse the long perspective of the commercial promenade through two diamond windows, transformed in the facade into an element of characterization of the pedestrian "portal".[4]

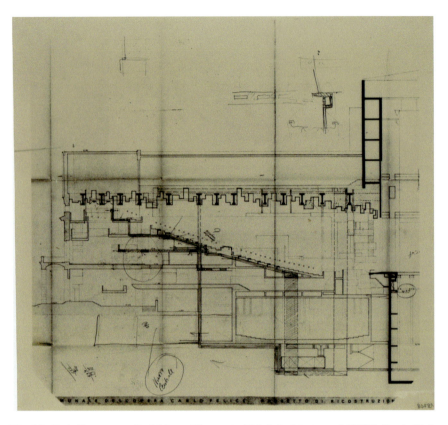

Fig. 6.5 Carlo Scarpa—studio del foyer [Centro Archivi di Architettura—MAXXI, Fondo Carlo Scarpa, Attività Professionale, Progetti e incarichi professionali, 214: Progetto Teatro Carlo Felice, Genova (1963–1981), n.18708]

In summary, Carlo Scarpa's project, in all subsequent re-workings and variations, hinges on three fundamental and important aspects: the continuation of the axis of Mazzini Gallery within the envelope of the new building, the conception of the foyer with facing corridors hanging on a common area and the opening of a large window that from the inside would have allowed to look towards the Gallery. By virtue of these insights, the foyer is made a real urban node, and the grafting of the new into the historical fabric can be said to be completely successful. The foyer is therefore transformed from a simple place of access and transit into a means of insertion, of penetration of the city into the theatre building; in it the hanging corridors like boxes would have allowed to observe the movement of the spectators before and after the shows. Scarpa develops the foyer in continuity with the floor directly above the external corridor coming from Mazzini Gallery. Its share is slightly higher than that of the stalls. The main purpose of the architect is therefore to connect the two levels with the third, represented by the main entrance to the theatre. Hence, there is the in-depth study of the connecting stairs and openings in the floors and along

6.1 A Theatre as an Urban Knot Grafted in the Historical Fabric

the paths, to favor visual and spatial continuity. In fact, the foyer creates a staging of the theatre itself, transferring, or even better, overturning the theatrical system of the large Italian hall, characterized by the typical visual contact, outside. However, it is difficult to understand with certainty whether this process was intended as a sort of "compensation" that Scarpa hypothesized, configuring the auditorium as a strongly unitary space, or even as a way of making everyday life spectacular.

About the hall itself: *[…] on one side it prepares a large internal spatial envelope that embraces the stalls, thus hiding the straight line of the wall structures; on the other, he works on the volumetric configuration of the proscenium.*[5] The aspect of primary importance that the latter plays in the architect's work should be emphasized. It is intended as a real junction point between two systems represented on one side by the stage, on the other by the theatre hall. Scarpa reinterprets one of the main and most recognizable characters of Italian theatre, giving a new guise and autonomy to the proscenium, a real aperture space.

It could be said that the image sought for the theatre by Scarpa is that of a "masonry" architecture. The composition generally takes place on the floors of the facades and, with the exception of some elements present on the view towards Mazzini Gallery, there seems to be no Scarpa's intention to decompose the building mass to make it deeper and three-dimensional. The main theme of the cladding is the alternation of stone bands of different thicknesses, which horizontally rhythm the views, reflecting the internal functional levels towards the outside. The design is intended to refer to the masonry textures of Genoese historical architecture, although it is unable to interpret the cadences and the structural sense, in the case of Scarpa reduced to a pure "graphic" texture of the covering. The linearity is interrupted by vertical windows, positioned in an irregular way, and by some elements of strong visual impact, such as the circular openings or the large low-arched window towards piazza Piccapietra. The only element that protrudes from the building volume is a long terrace that connects to the portal of the pedestrian passage of Mazzini Gallery, then joining the frieze of the Barabinian portico.[6]

The project presented in 1973 can be said to be a simplification of the previous one, both at a functional and a structural level. The foyer acquires the same axis as the main hall, with the stage located below the orchestra pit, becoming at the same time more capacious and easily accessible from the internal pedestrian tunnel.

Another substantial change is represented by the formal unification, in fact in a single architectural element, between the scenic tower and the space relating to the actual theatre hall, facing onto piazza Piccapietra.

The project report relating to the first version is reported in full in the apparatuses; in it, Andrea Palladio's Olympic Theatre in Vicenza, "*the most classic modern theatre for 'equal' spectators*" is defined as the main point of reference.

As mentioned above, contemporary critics have recognized a denial of Italian theatre—among which the main authors and reference projects include: Friedrich Gilly with the study for the Schauspielhaus hall in Berlin in 1798, Gottfried Semper with the Munich Opera in 1865 and the aforementioned Festspielhaus in Bayreuth in 1876—in the perspective concentration from the audience towards the space of representation as sense of organization of the theatre.

The approach to Palladio is not surprising, as we can rightly say that the two theatres in fact synthesize and make two antithetical conceptions of space for the show coexist, both between the ancient and the modern. Indeed, on the one hand, there is the hall for equal spectators, typical of the early theatre; on the other, the conception of the modern scene, with the city recreated in Palladio, with the staging towards the city in Scarpa.

Speaking of the negation of the Italian theatre is therefore, as mentioned above, to give a partial definition, since the characteristic system is in fact overturned in the foyer, integrated into the urban fabric.

6.2 From the Graft of the New to the Recovery of the Existing

In the following, we analyze some of the main Italian theatres, in the Italian type, which have determined the affirmation of the type, the favorable conclusion of the struggle for the formalization of the external aspect or rather of the autonomous building, of the representative organism placed in urban fabric. These great Italian achievements have at the same time been the subject of numerous interventions: from the grafting of the new, to the recovery of the existing, to conservation, up to the restoration and philological reconstruction. In the following insights, we wanted to keep a decreasing trend, starting from innovation towards conservation, in order to identify the reasons for the success of the type and the shared approaches to the theme, declining the concept of transformation.

The Teatro Regio di Torino[7] is a very illustrative case of the developing and characterization process of the architectural organism. The original project of one of the major eighteenth-century European theatres is due to Benedetto Alfieri, who in turn took up Juvarra's conception of Piazza Castello as a real business center. The Alfieri theatre, whose orientation was rotated ninety degrees with respect to the current theatre, was integrated into a larger building structure and still little architecturally denounced its presence, especially on the facade (Fig. 6.5). The representation of the project in the album of engravings *Il Nuovo Regio Teatro di Torino* of 1761 allowed the study and reconstruction of the first trial phase. In the aftermath of the unification of Italy, following ups and downs that had nevertheless seen substantially unchanged the plan, Italian type, with five orders of boxes, in 1865 the theatre was sold by the state property to the municipality due to the huge amounts needed for management. Therefore, the demand for democratization of public space took substance, with the need to replace the last two tiers of boxes with galleries. In particular, Donghi published in 1881 the *Progetto di un teatro notturno e diurno con annesso Salone e dipendenze ad uso del Municipio di Torino*. Finally, Ferdinando Cocito's project was approved and carried out over a period of twenty years; it accepts the request for democratization, although once again not configuring itself as an evolution or as a variation of the type. With the fire of 1936 and the demolitions of the building with the exception of the facade on piazza Castello, the formation phase of contemporary theatre began, based on the initial project by Morbelli and Morozzo, continued by Mollino until the inauguration in 1973. The requirements of the tender notice of the 1930s sought a mediation between the so-called reconstruction "as it was where it was" and the need for a modern building, which, however, would have followed the layout of the previous one, varying only the orientation of the stage. In the postwar period, however, the change took place. The new masterplan and the accessibility requirements of the building on the entire perimeter determine the variant of the rotation of ninety degrees of the axis of the theatre but, above all, its affirmation as an autonomous organism, which at the same time retains the characteristic integration with urban context. The entire building strongly moves back from the ancient facade, which is preserved and overturned, as well as the system of paths, in the foyer. The room becomes a nodal space, whose roof is a geometric, functional, constructive and esthetic synthesis. It is a cylindroid connected to a hyperbolic paraboloid in reinforced concrete that also includes the scenic tower, whose height remains low, respecting the limits of the context. Significantly, the solution was not produced by the individual designer (Musmeci was the person in charge of the structural design), but came from the construction practice, from the possibilities of the workers of the Bertone company. We can rightly speak of a long construction process and a continuous construction site, comparable to that of a medieval cathedral, which clearly defines and fixes the meaning of the developing process of the space for representation (Figs. 6.6 and 6.7).

Fig. 6.6 Main views of Turin drawn in perspective and carved in copper by the architect Giambattista Borra—The eighteenth-century theatre [https://www.teatroregio.torino.it/scopri-il-regio/storia]

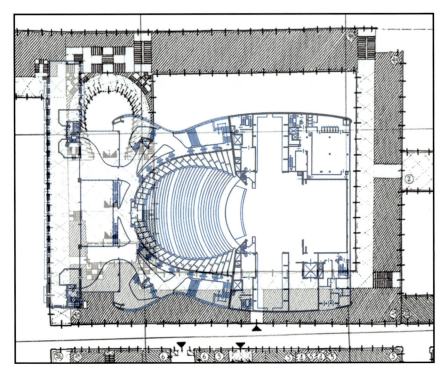

Fig. 6.7 The plan of the eighteenth-century theatre and that of the contemporary theatre by Carlo Mollino [elaboration by the author]

The Costanzi Theatre, later Opera di Roma, built in 1879, was conceived by the architect Sfondrini as a sound box, with a horseshoe shape, for an initial capacity of 2212 spectators, three tiers of boxes and gallery, surmounted from a dome. After the purchase by the Municipality in 1926, it underwent a partial restructuring entrusted to Marcello Piacentini, following which piazza Beniamino Gigli was created, the gallery was demolished internally and replaced by the fourth tier of boxes; in 1958, a new renovation led to the current facade with entrance and foyer. In the process of progressive acquisition of urban centrality, we therefore mention the unrealized project by Ludovico Quaroni, dating back to the 1980s. At the time, Purini reported about Piacentini's theatre: "*A facade perhaps designed for a larger size, built close to the theatre in the fifties on designs by Piacentini, finally collaborates in the exile of this open space in the crowded purgatory of urban interrupted architectures in Rome*", and again: "*urban architectures, Quaroni seems to suggest, to entrust their eternity to an external image and thus, in some way, to distance us from themselves, to exile us*".[8] Quaroni therefore aspired to "*do on the outside what has been done internally in several times, although with another spirit, perhaps with other means, with other, reduced competence*",[9] to make the building adequately representative of culture. The project was born to create a spacious foyer that acts as a hall of lost

6.2 From the Graft of the New to the Recovery of the Existing

steps; Quaroni, to give meaning to the ugly facades of the surrounding Umbertine Rome, proposes a large portico that cancels the square, consisting of 82 columns of pink granite, with industrial entablature in bronze or iron bars, resting on a stylobate of peperino, protruding ceiling above the entablature in pure reinforced concrete.

The 2002 project by Danilo Guerri and Paola Salmoni for Teatro delle Muse, on the other hand, brings back an exterior of the city inside, producing a true indoor replica of the Ancona urban facades. The neoclassical precedent of the theatre was represented by the Pietro Ghinelli work, built between 1824 and 1827 as an Italian hall with four tiers of boxes arranged in a horseshoe. It stood on an area known as the "Island of Prisons", a place where, in addition to the now dilapidated prisons, there were also the Finance building and some houses expropriated to be demolished. In particular, the block of San Filippo was demolished, which included the church and convent of the Filippini, as well as some private houses used as customs; the fountain of the Four Horses, originally located at the current portico of the Teatro delle Muse, was also moved. The rubble resulting from the demolition was reused for the construction of the Theatre, the Customs House, the Casino Dorico and the local houses, adjacent to the new structures. On the facade, the lower part of the current theatre consists of an ashlar porch of Istrian stone and the upper part of a sort of hexastyle Ionic temple, concluded by a balustrade and a continuous cornice.

The transformation undergone by the Teatro dei Filodrammatici is introverted too,[10] it is the first Patriotic Theatre, due to Luigi Caccia Dominioni in 1970 (Fig. 6.8). It is linked to the considerations made for academic theatres, understood

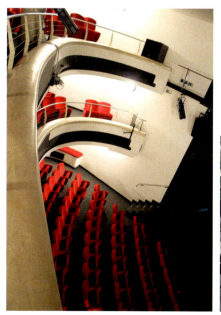

Fig. 6.8 Luigi Caccia Dominioni—Teatro dei Filodrammatici [photo by the author]

as synchronic variants of public box theatres. The theatre hall, located on the second basement floor with a capacity of 197 seats, is fluidly inserted in the multifunctional structure. It has two superimposed galleries, articulated as balconies given the limited spaces available; the curve is very flattened and shallow in the perpendicular direction to the soundstage. The main foyer is located on the floor of the stalls, equipped with a small bar and services, also designed by Luigi Caccia Dominioni. The facade of the ancient theatre, on the other hand, was maintained, rebuilt between 1905 and 1906 on a project by the architect Giachi, perfected by Laveni and Avati. The theatre, heavily damaged by the bombings of 1943, reopened three years later, thanks to the interventions of the architect Luigi Lorenzo Secchi. The adaptation of Caccia Dominioni was followed, just in 2015, by maintenance work and improvement of the stage.

From the projects described so far, it is clear that some of the major Italian theatres have come to the fore by virtue of substantial transformations and adjustments, which have changed their main characters and their relationship with the city, even on a symbolic-representative level. The new is grafted onto the Italian model, which has reached an urban as well as an architectural nodality in the proposed buildings. The strength and reason for the success of the type lies precisely in the great ability to absorb change. A shared approach was the interpretation of theatre as an event and a product of the transformations of the urban fabric, according to the interpretative line identified and proposed since the beginning of the research. The developing process of the modern space for the show inevitably tends towards continuous mutation into something else, it is based on previous intuitions to produce novelty; in this sense, speaking of a simple transformation of the existing can be limiting; the theatres

Fig. 6.9 Teatro Alla Scala [photo by the author]

6.2 From the Graft of the New to the Recovery of the Existing

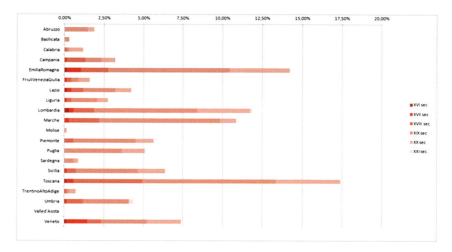

Fig. 6.10 Graph showing the distribution in time, by century of construction, and in space, by region, of the approximately 700 historic Italian theatres; note in particular the great development that took place in Emilia-Romagna, Lombardy, Veneto, Tuscany and Marche, in line with what has been described so far; the data are obtained from the "List of historical theatres in Italy", created as part of the research and reported in the appendix [elaboration by the author]

described so far can be said to be completely contemporary and different from the Italian theatre, although in continuity with it. The modern space for the show retains, as anticipated, a perspective orientation, with the stage distinct from the area intended for spectators, a democratization and a unification of the seats and environments of the theatre hall, a set of integrated paths and services, reproducing and overturning the city inside, in direct communication and correspondence with it (Fig. 6.10).

The subsequent case studies, on the other hand, concern, as anticipated, in a descending climax, theatres that have reached the Italian type system unaltered, often subject to effective maintenance interventions over time and/or conservation and recovery of the existing ones. The public theatre with boxes, it has been said, has known an unusual longevity in our country, being in fact still active, without any prejudice or almost for the conduct of the shows, the greatest number of historical theatres existing in Europe and in the West in genre. There are about seven hundred historical theatres in Italy, almost half of which were built in the nineteenth century, in accordance with the Italian type. We will see below how similar interventions can be innovative and indeed constitute a lifeblood for the buildings themselves.

Of the three major Italian theatres that are analyzed, La Scala, San Carlo and La Fenice, it is precisely La Scala (Fig. 6.9) that has undergone the most significant innovations in terms of distribution, which however concerned, beyond the scenic tower, almost exclusively the complex of the multifunctional building. Piermarini's 1776 project survives, although subjected to constant interventions,[11] almost unchanged in the layout of the hall. The Imperial Royal Architect and Inspector of factories for Lombardy completed the project of the Italian theatre par excellence in just 40 days, following the example of the Argentina and Regio theatres in Turin, respectively as regards the horseshoe shape, the stalls and the configuration of the service areas; the theatre has 4 tiers of boxes and 2 galleries (originally there were 5 orders to which the gallery was added), a large and sober foyer, a deep and functional stage, a majestic proscenium and an important system of annexes and services, which they aim to make the building autonomous. The same suffered, as it is known, serious damage from the bombing in the night between 15 and 16 August 1943, which led to urgent reconstruction works entrusted to Luigi Lorenzo Secchi from June 1945 to the spring of 1946, with the reopening concert directed by Toscanini. The following year the restoration of the museum began, in the 1950s Secchi also built the Piccola Scala and the corridor of the galleries. The continuous interventions of ordinary and extraordinary maintenance, plant adaptation and reconfiguration, if on the one hand they had had the merit of keeping the temple of opera alive, on the other they determined the need, no longer postponable, for more action on a wider scale. This is the meaning of Mario Botta's project for the new Scala starting from 2001 (Fig. 6.11), which gives legitimacy to the contemporary architectural language in its relationship with the historical monument. Botta limited himself to acting from scratch only in non-Piermarinian buildings, introducing two new backward volumes, one elliptical[12] and the other parallelepiped, respectively corresponding to the services and the scenic tower,[13] and creating in correspondence with the same two parallel axialities. The Ricordi casino itself had added the second axiality, recognized and strengthened by the new elliptical component. It has been said that Botta brought the tank back up to revive the historical image of him below. The scenic tower was demolished and rebuilt with the new machine designed by Fabio Malgrande; the hall, on the other

Fig. 6.11 Teatro Alla Scala [photo by the author]

6.2 From the Graft of the New to the Recovery of the Existing 143

hand, was subjected to purely conservative and restoration works, under the guidance of Elisabetta Fabbri; therefore a duplicity of approach for the two components of the theatre which will be discussed later, at the end of the chapter.

The San Carlo Theatre (Fig. 6.12) was born by the will of Carlo di Borbone, in order to give the capital of the Kingdom of Naples a theatre that would have replaced the ancient San Bartolomeo.[14] Inaugurated in 1737,[15] also to allow sufficient space behind the stage to the north of the Royal Palace and to the east of the Palazzo Vecchio, it involved numerous demolitions. It was built on the model of the Argentina Theatre, as well as the New Theatre of Montecalvario of 1724[16]; over time Fuga and Bibiena intervened, respectively to renovate the interiors and the acoustics; it was destroyed by a fire in 1816 and rebuilt by Niccolini according to the Medrano distribution, in a horseshoe shape with 6 tiers of boxes; already Niccolini inserted a set of functions into the complex, creating the foyer, the dance school and the scenography school; the portico for the transit of carriages and the loggia were also created, the lateral

Fig. 6.12 Teatro San Carlo [photo by the author]

access stairs were transformed into waiting rooms, and the connection with the Royal Palace was adapted. In 1843 Gavaudan and Gesuè won the competition for the reconstruction of the western facade of the theatre, following the demolition of the Palazzo Vecchio and the enlargement of the Royal Palace; other entrances were opened and the new northern wing of the Palace, rebuilt by the Genoeses, was made communicating, as well as with the theatre, with the Academy of the Knights. The royal box still has a large golden crown at the top. In 1872 the mystical gulf was created at the suggestion of Verdi, in 1890 electric lighting was introduced with the consequent abolition of the central chandelier, in 1927 the roof was completely rebuilt, in reinforced concrete, resting directly on the perimeter walls, in 1937 the new foyer and an adjoining side body were created. Following the bombings, the San Carlo resumed its regular activity in 1944; the set-up system and the carpentry in the basement were built; Subsequently, the reinforced concrete beams of the roof were replaced with new reticulars and, at the turn of the 90s, a substantial regulatory adjustment was made.[17]

The following describes, instead, in detail, the last major intervention undergone by the theatre, completed in 2010. In fact, unlike the Scala, the San Carlo was not involved in a project in the field of new, but related to the restoration and recovery of pre-existing buildings, under the banner of reversibility, authenticity and compatibility of the interventions, recognizing their stratification over time, which made it possible to interpret the theatre as a unitary complex. Restoration, rehabilitation, consolidation and re-functionalization have blurred boundaries between them, as evidenced by the designer, Elisabetta Fabbri. The main objective of the intervention on the building, as a UNESCO heritage site, was the increase of artistic production, to be achieved through the technological innovation of the stage, the construction of new rehearsal rooms, the adaptation of production sites and of the laboratories of Vigliena. It was therefore decided to fully preserve the external volumes, to add functions and services through low-impact changes, which would allow the continuity and unity of the architectural organism to be preserved. In place of the carpentries, transferred to the Vigliena laboratories, new rehearsal rooms have been created, a new transformation cabin has been built, which allows concerts to continue even in case of blackout without any interruption, a new hydronic air conditioning, as well as a Water Mist extinguishing system, and, following the verification of the building's seismic vulnerability, the roof was adapted with a new laminated wood structure[18], as well as the consolidation of the supporting walls. The hall itself has been made fully accessible to the handicapped, it has seen a series of conservative interventions of the decorative elements, but above all in it the stalls and the stage have been completely renovated, creating a new more functional slope for the former, and infinite combination possibilities for the latter thanks to the movable bridges; the interventions on the furnishings, and in particular on the upholstery, allowed an increase of 20% in the reverberation time, made necessary by the modern taste, to compensate the dryness of the sounds typical of stage theatres.

Teatro La Fenice (Fig. 6.13), as known, has undergone the most complete of interventions, the reconstruction "as it was where it was" carried out in 2003 by Aldo Rossi following the terrible fire of 1996. A project that in many respects coincides

6.2 From the Graft of the New to the Recovery of the Existing

Fig. 6.13 Teatro La Fenice—exterior [photo by the author]

with the approach described for the San Carlo, even if the damages reported by the Teatro La Fenice were such as to require, although appropriately mediated, the design of the new[19] (Fig. 6.14).

Finally, on a smaller scale for what concerns the size of the buildings affected by the interventions, but at the same very high level for the events, the example constituted by the two main theatres of Spoleto, the Caio Melisso and the Menotti Theatre (Fig. 6.15), is reported. The first, of seventeenth-century origin, is inserted, as well as the small Teatro delle 6 (Fig. 6.14), in the former Palazzo della Signoria, overlooking piazza Duomo; the second is instead located along the route that the architect Aleandri[20] had foreseen for the expansion of the city of Spoleto in the nineteenth century. Both theatres, in the Italian type, have undergone major conservative restorations and adaptations on the occasion of the establishment of the Festival dei Due Mondi, conceived by Gian Carlo Menotti 64 years ago. In line with what has been described so far, it is not surprising how the simple redefinition of the foyer of the Caio Melisso (Fig. 6.16) and the connecting stairs with the premises of the former Palazzo della Signoria, as well as the limited refurbishment of the attic and the vertical connections in the Menotti, in addition to the constant adjustments and maintenance, they alone were sufficient to allow an event of an international character, which covers the broad spectrum of modern theatre and arts. Furthermore, the spectacular vocation of piazza Duomo has already been mentioned.

Far from wanting to celebrate the ancient at all costs, we are witnessing, as anticipated, the current successful use of the Italian type and historical theatres in general. In our country, is it keeping what you have the only winning strategy?

Fig. 6.14 Teatro delle 6, ex Palazzo della Signoria in Spoleto [photo by the author]

Fig. 6.15 Teatro Menotti [photo by the author]

6.2 From the Graft of the New to the Recovery of the Existing

Fig. 6.16 Teatro Caio Melisso [photo by the author]

As it can be seen in Fig. 6.17, almost all of the historic theatres are buildings of clear architectural interest and/or listed, and over 10% of them are recognized as a World Heritage Site or included in a UNESCO site. This creates an "inevitable" demand in terms of conservative strategies. This is not the place to deepen these aspects, but certainly, for the purpose of researching the typological process of the modern space for the show, what emerges is the survival of the type thanks to the ability to welcome and integrate the new on the one hand, and to the possibilities of the technique on the other. This observation increases the value of the studio, as providing a new key to understanding modern space for the show that is, once again, the prerequisite for a correct design action.

6.3 Scenography as "Architecture of the Theatre Space"

In conclusion, the close link between Scenography and Architecture is mentioned; a scenography that is architecture from the moment in which it has historically determined and still determines the concretization of the theatrical building according to the constantly progressing spatial needs of representation. In this regard, it is reported: *"Overall, therefore, we note how the need for theatrical architecture to maintain its original intended use and to stage shows with modern technologies, although in a building strongly constrained by its own function and for this reason susceptible to not many changes from the architectural point of view, is the motivation that pushes to demolish and reconstruct the architectures of the historical scenic machine, or in any case, to strongly influence the design choices. [...] like the great opera houses, even for the architecture of more modest dimensions or, in any case, without a theatrical production, in a restoration intervention it is almost obvious the replacement of the staging apparatus in order to host a contemporary show, in view of economic as well as cultural purposes: it happens, therefore, that for the cultural good 'theatre' its being 'good' is considered prevalent over being a messenger of culture"*.[21] The conduct of extending also to structures of larger dimensions modest or with lower objectives, complete renovations, as seen precisely for Teatro alla

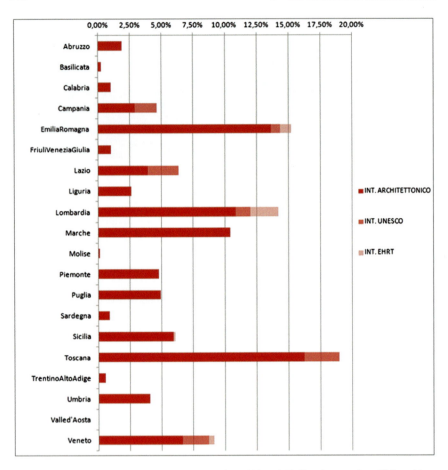

Fig. 6.17 Graph showing the distribution by region of historic Italian theatres for which architectural interest is recognized, of theatres that are World Heritage Sites or listed in Unesco sites and finally of theatres part of the European Route of Historic Theatres; the data are obtained from the "List of historical theatres in Italy", created as part of the research and reported in the appendix [elaboration by the author]

Scala, is not the only way to go. Alternative examples in this sense are the Misa Theatre of Ancona, inside Palazzo dei Priori, which has undergone an adaptation of the scenic tower translated into a distinguishable and reversible intervention; or the Court Theatre of the Royal Palace of Naples, in which a new scenic structure has been integrated into the more vulnerable elements of the existing one.

It is also true, on the other hand, that in a rapidly evolving environment such as the current one, in which a multiplicity of approaches is increasingly required, there is also the persistence, all Italian, of traditional scenography techniques, created to satisfy old spaces, designed in the past for past needs. Think, for example, of the laboratories of Teatro dell'Opera di Roma.[22]

6.3 Scenography as "Architecture of the Theatre Space" 149

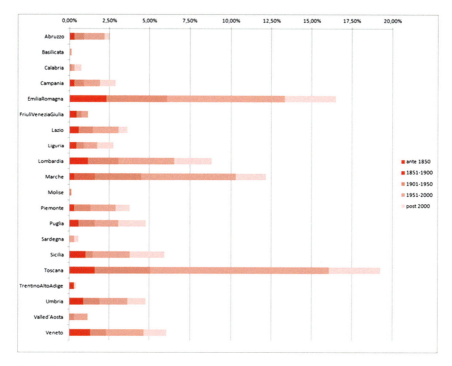

Fig. 6.18 Graph showing the distribution by region of historical Italian theatres subjected over time to interventions of grafting the new, recovery of the existing, conservation, restoration and philological reconstruction; the data are obtained from the "List of historical theatres in Italy", created as part of the research and reported in the appendix [elaboration by the author]

The issue is clearly very complex and therefore does not allow to be explored here. For the purposes of the research, however, it is important to underline the drive for transformation and continuous updating of the technique exerted by scenography and the performing arts in general. However, there is a need for a unified design action when these changes imposed by the surrounding conditions affect the spatial conformation, the architectural layout of the theatre building (Fig. 6.18).

6.4 Summary

We can say that we have investigated the evolution of the characters of Italian theatre by analyzing some significant case studies among recent projects.

It has been observed that, on the one hand, speaking exclusively of the transformation of the existing is limiting, as the developing process of the modern space for the show inevitably tends to continuous mutation into other, it is based on previous intuitions to produce novelty; and how, on the other hand, many redevelopment, recovery and restoration interventions constitute the lifeblood of existing buildings.

Finally, some of the main reasons for the success and consequent persistence of the Italian theatre typology were identified.

TEATRO CARLO FELICE — CODE6.01

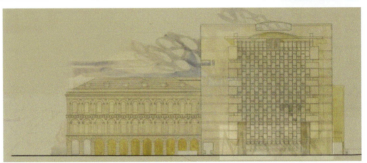

Carlo Scarpa - Teatro Carlo Felice di Genova - 1964

TYPOLOGICAL SCHEME

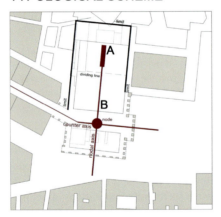

LEGEND

A STAGE
B AUDIENCE

SUMMARY OF TH MAIN SELECTION CHARACTERS

place_**GENOVA**
level of the event_**international**
site of architectural interest_**Yes (undone project)**
UNESCO site_**No**
year_**1964**
building_**permament theatre**
construction_**mixed structure**
features_**proscenium**
keyword_**urban node**

6.4 Summary

TEATRO CARLO FELICE — CODE6.21

design by BARABINO

- **stage:** □ flat, ■ inclined
- **scene:** □ fixed, ■ mobile — □ material, ■ perspective
- **scaenae frons:** ■ present, □ absent
- **proscenium:** ■ present, □ absent
- **separation audience/stage:** ■ present, ■ absent
- **audience:** □ "U" shape, □ "V" shape, □ jagged, □ bell shape, □ circular, □ oval, □ elliptical, □ double ellipse, ■ horseshoe, ■ mirrored — □ flat, ■ inclined
- **orchestra:** ■ bounded, □ not bounded — □ equal, ■ lower
- **theatre square:** □ present, ■ absent
- **seats:** □ with steps, □ with steps and loggias, □ with steps and boxes, ■ with boxes

design by SCARPA

- **stage:** □ flat, ■ inclined
- **scene:** □ fixed, ■ mobile — □ material, ■ perspective
- **scaenae frons:** □ present, ■ absent
- **proscenium:** ■ present, □ absent
- **separation audience/stage:** ■ present, □ absent
- **audience:** □ "U" shape, □ "V" shape, □ jagged, □ bell shape, □ circular, □ oval, □ elliptical, □ double ellipse, □ horseshoe, □ mirrored — □ flat, ■ inclined
- **orchestra:** ■ bounded, □ not bounded — □ equal, ■ lower
- **theatre square:** □ present, ■ absent
- **seats:** ■ with steps, □ with steps and loggias, □ with steps and boxes, □ with boxes

design by ROSSI

- **stage:** □ flat, ■ inclined
- **scene:** □ fixed, ■ mobile — □ material, ■ perspective
- **scaenae frons:** □ present, ■ absent
- **proscenium:** ■ present, □ absent
- **separation audience/stage:** ■ present, □ absent
- **audience:** □ "U" shape, □ "V" shape, □ jagged, □ bell shape, □ circular, □ oval, □ elliptical, □ double ellipse, □ horseshoe, □ mirrored — □ flat, ■ inclined
- **orchestra:** ■ bounded, □ not bounded — □ equal, ■ lower
- **theatre square:** ■ present, □ absent
- **seats:** □ with steps, ■ with steps and loggias, □ with steps and boxes, □ with boxes

152　　　　　　　　　　　　　　　　6　The Evolution of Theatre Features

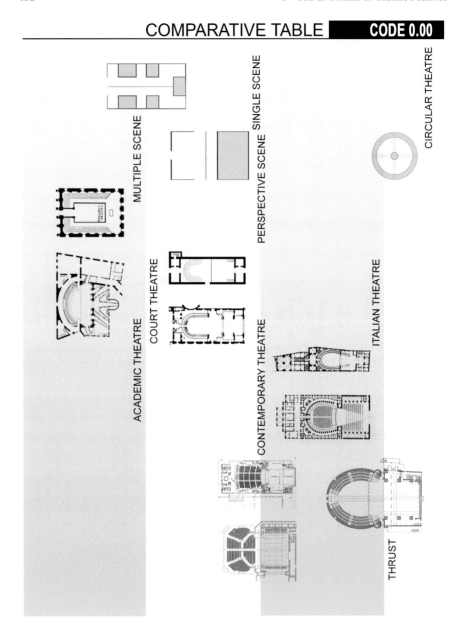

Notes

1. «*Le distruzioni belliche hanno reso pressoché inutilizzabili le strutture del teatro barabiniano. Il rilievo dello stato di fatto eseguito nel 1951 testimonia come siano resistite, sebbene fortemente danneggiate, le due facciate monumentali (il pronao—l'antico ingresso regio—verso piazza De Ferrari e il fronte*

verso via *XXV aprile), la struttura muraria della sala teatrale (che si sviluppa con la tradizionale pianta a campana) e alcuni spazi interni, in particolare il vestibolo dell'antico ingresso, con le otto colonne isolate che ritmano lo spazio tripartito. [...] Risultano invece inagibili la torre scenica, l'antico foyer collocato al primo livello sopra l'ingresso da* via *XXV aprile, come la maggior parte dei collegamenti verticali che conducevano gli spettatori ai livelli superiori dei palchi. [...] il teatro ha il compito di chiudere la monumentale cortina edilizia di piazza De Ferrari con il proprio fronte longitudinale. Barabino colloca infatti su questo fronte l'ingresso regio con il pronao caratterizzato da sei colonne doriche scanalate. Tuttavia questo affaccio non corrisponde all'asse del teatro, che trova invece il suo fronte naturale su* via *XXV aprile. [...] con il Piano Regolatore del 1954 si aggiungono nuove aspettative urbane all'edificio teatrale. Il Carlo Felice si trova, in pratica, in una condizione di snodo tra città antica, città ottocentesca (tra il grande asse di viale XX settembre e la Galleria Mazzini) e la nuova estensione del centro storico».* Valente E, Zanchettin V (2010) *I teatri di Carlo Scarpa.* Electa, Roma: 41–42

2. *«Esiste un progetto molto interessante di Carlo Scarpa—presentato molto prima del concorso a cui ho partecipato—che manteneva completamente il pronao per svilupparsi poi nel teatro vero e proprio e nelle sue parti laterali.* Rossi A (2002) Teatri, teatrini e spazi scenici». Malacarne G, Zimolo P M, Aldo Rossi e Venezia. Il teatro e la città. Edizioni Unicopli, Milano: 32

3. *«Il problema che mi sono posto era quello di cercare di incarnare, di assumere questa realtà storica, in una nuova realtà, in una nuova rappresentazione che avrebbe automaticamente tratto forza dagli elementi esistenti, quasi archeologici. [...] L'innovazione principale per ciò che riguarda la città è stata quella di creare, nella zona degli antichi ingressi, una continuità tra piazza De Ferrari e la Galleria Mazzini—oggi restaurata e ricca di negozi—che si trova dall'altra parte. Questo spazio di passaggio, è una sorta di piazza coperta in parte illuminata da una luce zenitale che arriva dall'alto mediante un elemento conico che attraversa e che illumina anche i vari foyer. Una delle caratteristiche del progetto è il superamento della classica pianta settecentesca a palchi, che forse per molti motivi non funziona più e che sarà riproposta invece nel progetto di ricostruzione del teatro La Fenice. Questo perché il palco è un'istituzione dell'aristocrazia ottocentesca che falsa notevolmente i calcoli relativi al numero degli spettatori in quanto, se in passato i proprietari li condividevano con guardie del corpo o figli giovani, oggi sono occupati solo da due o tre spettatori».* Ibidem: 32–34

4. *«Il pronao, oltre a fungere da ingresso all'edificio dalla piazza, diventa l'asse portante di un percorso pedonale urbano che attraversa il teatro e conduce fino al portale della galleria ottocentesca su piazzetta Labò.[...] Al centro del percorso trasversale dell'edificio, appartenente ancora al dominio del flusso pedonale della città, viene collocato l'ingresso al ridotto, che viene così completamente separato dal resto della macchina funzionale del teatro. Questa sembra essere la principale ragione per la quale i due ingressi monumentali al teatro vengono trasformati in snodi: da una parte entrambe servono*

l'atrio e il successivo sistema distributivo interno, dall'altra si collegano al sottostante prolungamento della galleria e alla sala piccola.[...] Prima di entrare nella sala, lo spettatore ha la possibilità di affacciarsi per l'ultima volta sulla Galleria Mazzini. In fondo all'atrio si sviluppa una curiosa rampa raddoppiata che porta inizialmente a un pianerottolo che fuoriesce dalla linea della facciata. Qui il visitatore ha la possibilità di scorgere la lunga prospettiva della promenade commerciale attraverso due finestre a diamante, trasformate nel prospetto in elemento di caratterizzazione del "portale" pedonale». Valente E, Zanchettin V (2010) *I teatri di Carlo Scarpa.* Electa, Roma: 42–43

5. *«[...] da una parte predispone un grande involucro spaziale interno che abbraccia la platea, nascondendo così l'andamento retto delle strutture murarie; dall'altra lavora sulla configurazione volumetrica del proscenio».* Ibidem: 44–46

6. *«Si potrebbe affermare che l'immagine ricercata da Scarpa per il teatro sia quella di un'architettura "muraria". La composizione si svolge generalmente sui piani delle facciate e, salvo alcuni elementi presenti sull'affaccio verso la Galleria Mazzini, non sembra esserci alcuna intenzione da parte di Scarpa di scomporre il volume della massa edilizia per renderla più profonda e tridimensionale. Il principale tema del rivestimento è l'avvicendamento di fasce di pietra di diverso spessore, che ritmano orizzontalmente gli affacci, riflettendo verso l'esterno i livelli funzionali interni. Il disegno vuole rimandare alle trame murarie dell'architettura storica genovese, sebbene non riesca a interpretarne le cadenze e il senso strutturale, nel caso di Scarpa ridotto a pura trama "grafica" di rivestimento. La linearità è interrotta da finestre verticali, posizionate in maniera irregolare, e da alcuni elementi di forte impatto visivo, come le aperture circolari o la grande finestra ad arco ribassato verso piazza Piccapietra. L'unico elemento che sporge dal volume edilizio è un lungo terrazzo che si allaccia al portale del passaggio pedonale della Galleria Mazzini, legandosi poi al fregio del portico barabiniano».* Ibidem: 46—48

7. The Teatro Regio had a precedent in the Teatro delle Feste integrated into the old Ducal Palace in the last twenty years of the seventeenth century, it was a precarious wooden structure, which caught fire and disappeared already during the eighteenth century, replaced by the Teatrino del Rondò.

8. *«Una facciata forse pensata per una dimensione maggiore, costruita a ridosso del teatro negli anni cinquanta su disegni dello studio Piacentini collabora infine all'esilio di questo slargo nel purgatorio affollato delle architetture urbane interrotte di Roma",* and: *"le architetture urbane, sembra volerci suggerire Quaroni, affidano la propria eternità ad un'immagine esterna e così, in qualche modo, ci allontanano da se stesse, ci esiliano».* Purini F (1983) Osservazioni sul complesso dell'Opera di Roma di Ludovico Quaroni. In: Lotus (4): 29–35

9. *«fare all'esterno quanto in più tempi è stato fatto all'interno, sia pure con altro spirito, forse con altri mezzi, con altra, ridotta competenza».* Ibidem

10. The Patriotic Theatre was established on the ancient church of Saints Cosma and Damiano, then Palazzo Vimercati, of which the fifteenth-century portal is

6.4 Summary

still preserved today. Canonica, who received the church in 1799, suggested not to demolish it but to adapt it. A design by Piermarini and the proclamation of the second Cisalpine Republic made it possible to resume work on the lot to be used as a theatre. Gaetano Vaccani restored the decorative apparatus in 1801. The years between 1849 and 1851 marked the adaptation and reopening of one of the most elegant theatres in the city.

11. In 1807, the theatre was internally renovated for the first time; between 1813 and 1814, there was the enlargement of the stage and the consequent demolition of the adjacent buildings. In 1823 the Sanquirico chandelier was inaugurated; in 1830, the two lateral bodies crowned by terraces were added, the velarium and the decorative apparatus were reconstituted. In 1845, the reconstruction work on the roof of the old painters' hall was carried out; in 1860, gas lighting was introduced, modeled on the Phoenix. In 1883, it was equipped with electric light. In 1891, the fifth tier of boxes was abolished and transformed into a gallery, the flow of the public was regulated and the seats reduced, as well as the standing ones abolished. Between 1891 and 1907, the orchestra pit was lowered, between 1891 and 1909, the gallery was transformed into a second gallery equipped with a foyer, bar and services. In 1910, the architect Tononi designed the Teatro alla Scala Museum, inaugurated in 1913, communicating with the reduced. In 1920, there was a new modernization of the stage; in 1921, the interior furnishings of the boxes were unified; the third order in particular included the large foyer, a series of smaller foyers and the terrace; and, finally, in 1937, the movable bridges were built.

12. The elliptical volume rises where the Piccola Scala once stood.

13. Respectful of the heights reached by the previous turrets built in the 1930s.

14. The San Bartolomeo Theatre was built in 1620 under the Viceroy Ognatte, in the area of the current Santa Maria delle Grazie; it had a traditional rectangular plan, 34 boxes on 2 orders, gallery and stalls with 300 seats. It was destroyed by a fire in 1681 and rebuilt in 1684. The King bought it to have it demolished and reused the timber for the construction of the San Carlo.

15. The ancient theatre had a single-storey building which included the three stairways and the side door on the street, an adjacent volume on two floors containing the entrance corridors to the stalls and the royal box, and a larger body comprising the six tiers of boxes, the lateral access stairs, the stalls and the stage; the trussed roof insisted both on the hall and on the stage, the height of the stalls was raised by 3 m with respect to street level; finally, there was a body at the lower end, behind the stage, used as a scene depot and three side rooms for props and wardrobe. The design of a 1747 scene by Vincenzo Re for the birth of the heir to the throne Philip has allowed us to trace the original configuration of the theatre.

16. Made on a project by Antonio Domenico Vaccaro, who introduced the horseshoe plant in Campania.

17. Among the adaptation interventions carried out, the following should be noted: the consolidation of the roof beams of the stage and of the supporting structures of the stages, the upgrading of the electrical and smoke detection system, the

restoration of the underground fire-fighting tank, the construction of elevators accessible to disabled people, the recovery of the premises below piazza San Ferdinando, the reconstruction of the set-up apparatus and the consolidation of the wooden structures of the understages, the restoration of the floors and upholstery with fireproof materials, the redevelopment of the administrative offices.

18. The main structure of the roof consists of 4 trusses in glued laminated wood, with rafters with curved intrados and straight extrados, resting on a reinforced concrete curb and mutually connected by orthogonal struts, 5 movable, tiltable bridges, with maximum excursion $+2$, 5 to 2 m.

19. In 1789, following a competition notice containing the map of the area and 14 points to be followed for the design of the new theatre in Venice, Stratico, Buratti and Fontanesi selected the Selva project. This was completed in 27 months; it included a horseshoe plan, 5 rows of 35 boxes, a Corinthian tetrastyle loggia on the facade, five ashlar arches on the opposite front. A fire destroyed it in 1836; it was rebuilt in 1837 by Meduna as per the Selva project.

20. Ireneo Aleandri, in charge of the project for the New Theatre in 1853, created the building as a real urban monument, with a neoclassical facade including three orders, an advanced ashlar portico, a colonnaded atrium connected to the side rooms. The layout is horseshoe-shaped, with 4 orders of 23 boxes each and a gallery; the so-called Vanvitellian umbrella constitutes the attack of the gallery on the ceiling; there are also a lower wing containing services and two internal courtyards.

21. *«Nel complesso, dunque, si nota come la necessità dell'architettura teatrale di mantenere la destinazione d'uso originaria e di mettere in scena spettacoli con tecnologie moderne, seppur in un edificio fortemente vincolato dalla sua stessa funzione e per questo suscettibile di non molte mutazioni dal punto di vista architettonico, sia la motivazione che spinge a demolire e ricostruire le architetture della macchina scenica storica, o comunque, ad influenzare fortemente le scelte progettuali. [...] al pari dei grandi teatri lirici, anche per le architetture di dimensioni più modeste o, comunque, prive di una produzione teatrale è quasi scontata, in un intervento di restauro, la sostituzione dell'apparato scenotecnico affinché si possa ospitare uno spettacolo contemporaneo, in vista di finalità economiche oltre che culturali: accade, quindi, che per il bene culturale 'teatro' sia considerato prevalente il suo essere 'bene' sull'essere messaggero di cultura».* Bortolotti L, Masetti Bittelli L (1997) Teatri storici. Dal restauro allo spettacolo. Nardini Editore, Fiesole (FI)

22. See also Clemente S (2015) "La fabbrica dei sogni". In: *Memorabilia. Nel paese delle ultime cose.* Aracne Editrice, Roma

References

(2004) La nuova Scala. Marsilio Editori, Venezia
(2010) Il Teatro Regio di Torino da Carlo Mollino a oggi. Dario Flaccovio Editore, Palermo
(2010) Teatro di San Carlo. Memoria e innovazione. Arte'm, Napoli
(2015) Il Teatro di San Carlo Napoli. Sagep Editori, Genova
Augenti N (2006) Storia e vicende edilizie del Teatro di San Carlo a Napoli. In: Buccaro A, Fabricatore G, Papa LM (ed) Storia dell'ingegneria. Cuzzolin Editore, Napoli
Bortolotti L, Masetti Bitelli L (1997) Teatri storici. Dal restauro allo spettacolo. Nardini Editore, Fiesole (FI)
Botta M, Fabbri E, Malgrande F (2005) Il Teatro alla Scala. Restauro e ristrutturazione. Skira, Milano
Capella M (2007) Il teatro degli artisti, da Picasso a Calder, da De Chirico a Guttuso, scene, bozzetti e costumi dal Teatro dell'Opera di Roma. Silvana Editoriale, Milano
Capici GC (2003) Il Teatro Nuovo di Spoleto. Pilaedit, Roma
Fabbri P, Bertieri MC (2004) I teatri di Ferrara. Il Comunale. LIM Editrice, Lucca
Guicciardi E (1970) Il nuovo teatro di un'accademia milanese 1798–1970. Accademia dei Filodrammatici, Milano
Menozzi L (1983) Ludovico Quaroni. Il progetto per il Teatro dell'Opera di Roma. Controspazio (4):106–112
Moschini F (1983) Dal dubbio alla certezza. Un progetto di Ludovico Quaroni. Lotus (40):23–28
Orlandi A (1986) Ludovico Quaroni: dieci quesiti e cinque progetti. Officina Edizioni, Roma
Purini F (1983) Osservazioni sul complesso dell'Opera di Roma di Ludovico Quaroni. Lotus (40):29–35
Strappa G (2014) L'architettura come processo. Il mondo plastico murario in divenire. Franco Angeli, Milano
Strappa G, Carlotti P, Camiz A (2016) Urban Morphology and Historical Fabrics. Contemporary design of small towns in Latium. Gangemi Editore, Roma
Tosti Croce M, Ciarlantini P (2011) La fabbrica delle meraviglie. il lavoro editoriale, Ancona
Valente E, Zanchettin V (2010) I teatri di Carlo Scarpa. Electa, Roma

Documents

SERIE: PUBBLICHE COSTRUZIONI Milano, città (P), Como, città Como, provincia (P) Pavia, città (P), Pavia, provincia (P) 1821–1832, Ferranti (ingegnere), pareri, Milano, teatro alla Scala; Pavia, Colombina, carceri, ICONOGRAFIE: Scala, prospetto palchi - 1830 (disegno), Scala, palchi, sezione - 1830 (disegno), Scala, profilo nuova cornice (disegno), Scala, particolari costruttivi (schizzi 5) [Archivio di Stato di Milano, Genio Civile di Milano, busta n. 3666]
SERIE: PUBBLICHE COSTRUZIONI Milano, città (P), 1849–1853, Scala (teatro alla), lavori nel teatro alla Scala, sezione trasversale del teatro La Fenice in Venezia, dimostrante il sistema d'illuminazione - 1845 (disegno), becco di fiamma a ventaglio - 1845 (disegno), drappeggi di stoffa (disegni acquerellati), piante dei piani del casino annesso alla Scala, rilegati in fascicolo - 1850 (disegni), ICONOGRAFIE: "nuova copertura sul salone vecchio dei Pittori e sale annesse..." [Archivio di Stato di Milano, Genio Civile di Milano, busta n. 3679]
SERIE: PUBBLICHE COSTRUZIONI Milano, città (P), 1848–1852, Scala (teatro alla) Cannobiana (teatro alla) [Archivio di Stato di Milano, Genio Civile di Milano, busta n. 3680]
SERIE: PUBBLICHE COSTRUZIONI Milano, città (P) 1845–1854, Scala (teatro alla) Scala (casino presso la) Cannobiana (teatro alla), illuminazione a gas, illuminazione a gas della Scala; fornitura di tele per le quinte di Scala e Cannobiana, casino presso la Scala, memorie riguardanti

l'affitto Ricordi; "Consegna del Casino Erariale... a Giovanni Ricordi editore di Musica", fascicolo rilegato con piante dei piani [Archivio di Stato di Milano, Genio Civile di Milano, busta n. 3840]

SERIE: FABBRICHE ERARIALI Milano, città (P) 1817–1821, Cannobiana (teatro) Scala (teatro), inventari [Archivio di Stato di Milano, Genio Civile di Milano, busta n. 4303]

SERIE: FABBRICHE ERARIALI Milano, città (P) 1816, Cannobiana (teatro) Scala (teatro), consegne e riconsegne [Archivio di Stato di Milano, Genio Civile di Milano, busta n. 4827]

SERIE: CORPO REALE DEL GENIO CIVILE, FABBRICATI Milano, città (P) 1868–1879, Scala (teatro alla) Arena (anfiteatro) San Marco (magazzino del sale) Milano—amministrazione comunale di, cessione di beni demaniali al Comune di Milano: casini alle porte, inventario del teatro alla Scala per la consegna all'impresa Bonola e la riconsegna al Comune, consegna dell'ex magazzino centrale del Sale in San Marco all'acquirente Comune di Milano, cessione al Comune dell'anfiteatro Arena, ICONOGRAFIE: corsi d'acqua esistenti nell'area dell'anfiteatro—1872 (pianta), teatro alla Scala—1832 (pianta) [Archivio di Stato di Milano, Genio Civile di Milano, busta n. 6391]

SERIE: CORPO REALE DEL GENIO CIVILE, FABBRICATI Milano, città (P), 1856–1871, Scala (teatro alla) Cannobiana (teatro alla), affitto del casino Ricordi; teatro alla Scala, coerenza Valentini; personale della Scala; "macchinismo Ronchi, onde figurare il tuono, il lampo, la folgore, la grandine e la pioggia in servizio dell'I. R.Teatro alla Scala", ICONOGRAFIE: Scala, coerenza Valentini - 1856 (planimetrie 3), tavole rilegate del "macchinismo Ronchi"—1856 (disegni) [Archivio di Stato di Milano, Genio Civile di Milano, busta n. 6401]

Milano—Filodrammatici e Patriottico [Archivio di Stato di Milano, Atti di governo, Spettacoli Pubblici parte moderna, busta n. 37]

Fenice Teatro [Archivio di Stato di Venezia, Governo Veneto—Atti—Atti 1822—CXLV teatri e spettacoli, busta n. 2362]

Teatri—Comuni—Milano—Carcano [Archivio di Stato di Milano, Spettacoli Pubblici Parte Moderna, busta n. 38]

Teatri—Comuni—Milano—P.G.—Teatri diversi [Archivio di Stato di Milano, Spettacoli Pubblici Parte Moderna, busta n. 39]

Teatri—Comuni—Milano—Scala—Canobiana—P.G. [Archivio di Stato di Milano, Spettacoli Pubblici Parte Moderna, busta n. 40]

Teatri—Comuni—Milano—Re [Archivio di Stato di Milano, Spettacoli Pubblici Parte Moderna, busta n. 42]

Teatri – Comuni—Venezia—Fenice—P.G.—Appalti [Archivio di Stato di Milano, Spettacoli Pubblici Parte Moderna, busta n. 54]

Progetto Teatro Carlo Felice, Genova (1963–1981) [Centro Archivi di Architettura—MAXXI, Fondo Carlo Scarpa, Attività Professionale, Progetti e incarichi professionali, 214: 18576, 18645, 18673R, 18701R, 18708, 18776]

Conclusions

In the presented research, an interpretative line of the theatrical organism was outlined, which recognizes its derivation from the transformations of the fabric of our cities, parts of a constant process.

In particular, having defined the space for the show as a changeable urban event, the distance from the model of classical theatre was understood and overcome, recognizing an essential function of innovation to the activities developed throughout the entire medieval period, undoubtedly the most complex to interpret due to the scarcity of sources, and of which some attempts at reconstruction have been proposed.

The focus was therefore placed on primary construction actions, characterized by a minimum level of typicality, of fencing and covering, at the base of the process. The simplest form, as we have seen, is that of the fence, the courtyard area, the square or, more generally, the city, which is identified as suitable for fulfilling the special, temporary, theatrical function.

Through some case studies placed primarily within the courts and academies, the story of the Municipal Palace of Ferrara is remembered in particular, but also in the contemporary world, outside a purely chronological order and instead within a constant process, the characterization of the modern space for the show is described. In particular, it was emphasized how temporary forms of space organization were precursors of some of the main characters of modern theatre.

With the *inventiones*, original contributions to the formation of the type, by Palladio, Scamozzi, Aleotti and many other authors up to the present day, it was possible to solidify the organic and unity vocation and reach the full autonomy of the theatrical building, with their own permanent, specific distinctive signs.

It was then demonstrated, starting from some partly unpublished episodes in the city of Mantua, how the research and mutation of the urban space used as a theatre has been and is still constant. Thanks to the contribution of Carini Motta and his *Trattato sopra la struttura de' theatri, e scene*, the uniqueness of the developing process of modern theatre has been proven, by progressive specialization, starting from the basic building as house is, form of the simplest way of living, or as building, already at a later level, towards the same specialization. The descriptions of the court

---

© Springer Nature Switzerland AG 2022
S. Clemente, *Reconstructing Theatre Architecture*, The Urban Book Series,
https://doi.org/10.1007/978-3-030-89968-4

theatre and the public theatre with boxes, the germ of the Italian theatre, correspond to the two starting points, actually of equal significance.

In summary, we can say that we have investigated the development and evolution of the characters, analyzing national projects up to the recent era, identifying the main reasons for success and affirmation and mentioning the main influences, incoming and outgoing, of the Italian type at a European level.

Since the perfect parity existing in the architectural project between reading and writing, we hope, as a future perspective, the interpretative line provided could be useful for understanding and proposing the transformations of the existing heritage, as well as for the creation of the new one.

The contemporary intervention is grafted onto the Italian model, which has reached an urban as well as an architectural nodality in many of the described buildings. The strength and reason for the success of the type is precisely the great ability to absorb change. The developing process of the modern space for the show inevitably tends towards continuous mutation into something else, it is based on previous intuitions to produce novelty; in this sense, these theatres can be said to be completely contemporary and different from the Italian-type theatre, though in continuity with it.

A further trend is represented by theatres that have arrived unaltered in the system, often subject of effective maintenance interventions over time and/or conservation and recovery of the existing one. The public theatre with boxes, it was said, has known an unusual longevity in Italy, being in fact still active, without any prejudice or almost for the shows, the greatest number of historical theatres existing in Europe and in the West in kind. Almost all of these are buildings of clear architectural interest, in some cases, recognized as World Heritage Sites. Especially in this unique context, the reading of the process appears essential.

In support of the aforementioned perspectives, we also add the importance of the drive to transform and continually update the technique, exercised by scenography and performing arts in general, for which we need to understand the process and to consequently coordinate a unitary project.

We hope that the conducted analysis could be useful in orienting the design research of the new and new places in the city for the modern space for entertainment, constantly active, as represented by the Milanese example.

Conclusions

161

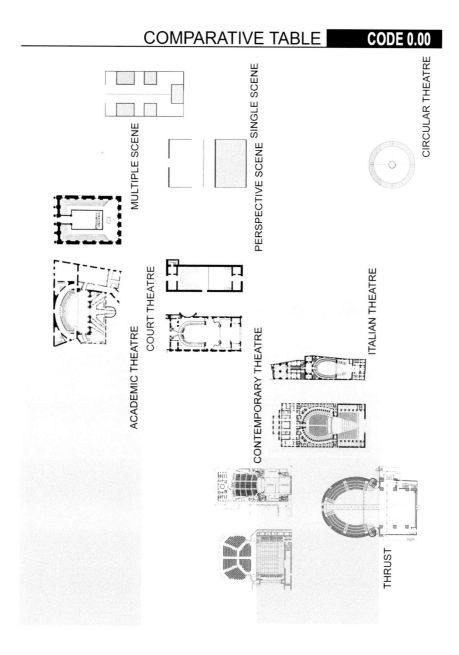

Apparatuses

Literature

From Fiocco G (1958) La casa di Alvise Cornaro. Storia e Letteratura (72): 69–77.

Pratico di architettura fu dunque il Cornaro; uomo di gusto, non architetto, e sembra si possa far entrare appunto in tale nobile pratica quel tanto della sua dimora padovana che non assunse aspetti monumentali, ma soltanto quelli, per intenderci, di ordinaria amministrazione. Nessuno se l'è chiesto, perché l'imponenza dell'Odeon e della Loggia hanno polarizzato l'attenzione degli studiosi e degli uomini di gusto. Ma essi non bastavano per fare la casa del Cornaro. Bisognava rendersi conto della facciata, prospettante l'odierna via Cesarotti e dell'ala che, contrapponendosi all'Odeon, e legandosi alla famosa Loggia dove il Ruzzante recitava, compiva la precinzione dell'ampio cortile dell'abitazione di Alvise […].

È questo aspetto che ho voluto innanzi tutto richiamare, per vedere, in un secondo tempo, se potesse convenire al Cornaro. A ciò mi ha giovato una preziosa stampa che accompagna, con altre belle vedute, la «pianta di Padova», incisa da G. Valle nel 1784. Essa prova che Odeon e Loggia non erano i soli edifizi del vasto cortile, posto dietro alla casa propriamente detta; esistevano costruzioni a logge; semplici ma decorosissime, e raccordi di muraglie e di passaggi, in armonia con le architetture del Falconetto, col quale facevano un complesso preciso e bilanciato, senza meticolosità, ma con grazia. Per il prospetto sulla via Cesarotti siamo meno fortunati, ma non sprovveduti di un qualche documento. Senza la maestria certo del Valle, ma con sufficiente precisione, Lorenzo Mazzi, pubblico perito, l'ha tratteggiata nel 1735 nei volumi che documentano i livelli e le proprietà spettanti in Padova al Convento di Sant'Antonio. È un disegno corsivo, spoglio di pretese, ma sufficiente per attestare nella balconata centrale, di tipo lombardesco, una certa corrispondenza con la loggia del fabbricato, a sinistra della corte che sta dietro. La facciata, negli altri balconi obliterata da chiusure lignee, aggiunte per attutire i rigori invernali, aveva del resto, in origine, altro aspetto, in quanto era stata decorata da affreschi, «more venetiano», da Gerolamo de' Sordi, detto del Santo per il luogo della abitazione, prossima alla

© Springer Nature Switzerland AG 2022
S. Clemente, *Reconstructing Theatre Architecture*, The Urban Book Series,
https://doi.org/10.1007/978-3-030-89968-4

famosa chiesa e prossima alla casa Cornaro, come ha documentato E. Rigoni nel suo nutrito saggio intorno al maestro, edito nel 1941 nelle Memorie dell'Accademia di Padova (vol. 57, doc. VII) [...].

Vicinanza che ha certo favorito l'affrescatura, di cui posso dar notizia qui per la prima volta. Tutti questi edifici di comodo potrebbero essere assegnati al Cornaro stesso, sebbene non si abbia modo di stabilirlo. Pure ammettendo che per essi non sia possibile invocare il nome del Falconetto. D'altra parte potrebbe darsi che la casa propriamente detta, prospettante su via Cesarotti, fosse il frutto di rimaneggiamenti, specie nelle ali, tanto asimmetriche e senza significato [...].

Così l'aspetto della casa Cornaro ci è fatto cognito nel suo complesso; con non piccolo vantaggio anche per la valutazione dell'opera insigne del Falconetto, la quale evidentemente, come ogni costruzione di buona architettura, dovette tener conto delle fabbriche esistenti. Ne nacque un complesso, non solo timbrato da due edifici eccezionali, ma inteso come un «girotondo», dove il meno dà forza al più, allacciandosi armonicamente, a braccia stese, a mani conserte, per formare un tutto d'innegabile eleganza anche là dove è in tono minore [...].

Finora le due eccezionali fabbriche dell'Odeon e della Loggia erano sembrate tutt'al più le parti di un mondo architettonico incompiuto; mentre ne sono i resti preziosi. Restaurare l'antico aspetto, almeno nella memoria e nella testimonianza dimenticata, spero non sarà parso inutile a chi ama il rinascimento appunto per questa civiltà costruttiva, che rispetta non solo se stessa, la propria voce cioè, per quanto impegnata sia, ma anche quella degli altri, ed accordandosi coralmente avvantaggia sé nel tutto.

From Tamassia LO (1999) *L'edificio del mercato dei bozzoli. La storia del sito.* **Quaderni di Archeologia del Mantovano Museo Civico Archeologico di Ostiglia Gruppo Archeologico Ostigliese (1): 191–223.**

Il Mercato dei Bozzoli (1896–1930).

(a)   Le demolizioni e i progetti

Ampia documentazione ottocentesca, conservata nell'Archivio di Stato e nell'Archivio Storico del Comune di Mantova, consente da questo momento in poi una puntuale ricostruzione della vicenda dell'alienazione e della demolizione del Teatro Regio. Vicenda penosa ai nostri occhi, avendo la comunità mantovana perduto irrimediabilmente una copia originale, benché ridotta ed adattata, del prestigioso teatro della Scala di Milano: ma la storia segue il corso e le carte non fanno che riportare fedelmente i fatti.

Apprendiamo dunque che una locale "Commissione speciale per la sistemazione del Palazzo ex Ducale di Mantova", appositamente istituita dal Prefetto, interpellata dal Ministero della Pubblica Istruzione circa l'opportunità di alienare o meno il teatro di Corte, nel contesto di un generale ripristino del complesso Ducale, così si pronunciò nella seduta del 16 ottobre 1886: "La Commissione unanime riconosce che nel fabbricato del teatro nulla vi è di monumentale né di storico-artistico e perciò opina che debba essere venduto ... possibilmente al Comune, all'effetto di scoprire anche dal lato della città il bellissimo Castello".

Apparatuses 165

Così nonostante dal fitto carteggio successivamente intercorso con il ministero risulti una momentanea sospensione della decisone, causa il parere negativo della "Commissione Provinciale per la conservazione dei Monumenti", la sentenza era già stata pronunciata: l'edificio sarebbe stato immolato all'imperante gusto archeologico-medievalistico del momento, che faceva prediligere i monumenti di sapore medievale alle costruzioni dei secoli successivi.

Altra vittima illustre della medesima congiuntura culturale sarà infatti la giuliesca palazzina della Paleologa, che pure vediamo rappresentata nel rilievo della "pianta terrena del Reale Palazzo in Mantova", prodotto a corredo della documentazione di cui sopra.

E' del 12 gennaio 1887 la comunicazione da parte del Ministero della Pubblica Istruzione della decisione finale di consentire la vendita del teatro, a condizione "che l'acquirente sia il Comune... e che il Teatro sia demolito per migliorare le condizioni della porta di S. Giorgio e per mettere in maggiore evidenza il Castello".

L'atto di compravendita del fabbricato è datato 24 marzo 1896; dalle notizie in esso contenute apprendiamo che l'acquisto fu autorizzato con verbale del Consiglio Comunale del 5 aprile e 5 luglio 1895, approvato dalla Giunta Provinciale Amministrativa il 13 agosto del medesimo anno. Il prezzo convenuto fu di L. 6750 ed il contratto previde alcune "condizioni speciali per la cessione" visivamente evidenziate nel disegno allegato all'atto: mentre il voltone d'ingresso a piazza Castello ed il portico della stessa lungo il lato nord dovevano essere mantenuti inalterati, in quanto parti integranti della Reggia, il Comune doveva provvedere all'immediato abbattimento di alcune parti esterne al corpo principale. Si trattava di una superfetazione già pericolante, nata presumibilmente nel corso dell'Ottocento lungo via S. Giorgio, per contenere i camerini addetti ai palchi del lato destro del teatro, e della sala del ridotto con i locali annessi, per restituire alla vista il torrione nord del Castello.

Nel corso del 1897 si avviò la discussione intorno alle possibilità di utilizzo dello stabile: varie proposte di adattamento e relativi preventivi furono redatti dall'ingegnere municipale Carlo Andreani, padre del più famoso Aldo, l'architetto autore dei progetti di importanti edifici di Mantova, tra cui la sede della Camera di Commercio. Risalgono ai mesi di settembre e ottobre di quell'anno i primi progetti e le relative stime: la copiosa documentazione conservata all'Archivio Storico del Comune comprendente anche vari rilievi su lucido del teatro, è utile non solo per ripercorrere le fasi progettuali del Mercato dei Bozzoli, ma anche per conoscere, quasi in controluce, le caratteristiche strutturali e logistiche del grande fabbricato che, come abbiamo visto, aveva ospitato i teatri di corte. Apprendiamo infatti, da una dettagliata "Relazione sulle condizioni di stabilità del Teatro Regio" che l'edificio, "costruito con antichi sistemi per distribuzione per ordini di logge e di palchi, con un solo ingresso a mezzo di stretto corridoio lungo m. 36, non poteva più servire per uso di spettacoli, date le esigenze moderne dopo i recenti disastri di teatri". Ma punto dolente del grande fabbricato era il tetto, che, formato da grosse capriate lunghe circa 18 metri, e con pendenza del 60%, era responsabile, causa lo scorrimento delle tegole verso il basso, di diffuse infiltrazioni d'acqua piovana e conseguente infradiciamento delle sottostanti travi lignee e danneggiamento del soffitto del palcoscenico, platea e loggione. La relazione dell'Andreani sul grave dissesto statico dell'edificio prosegue con l'analisi delle tecniche costruttive e con la messa in esistenza dei conseguenti difetti strutturali per giungere alla conclusione che era urgente "demolire tutto il teatro, conservando di esso solo i muri di perimetro, perché, tranne alcune tracce di crepacci, dovute a cedimenti del sottosuolo, non presentano altri inconvenienti di stabilità".

Un importante disegno su lucido, datato 6 ottobre 1897 a firma di Carlo Andreani, evidenziando cromaticamente le operazioni da compiere (in giallo le parti della cavea da demolire, in rosso i pilastri da costruire) consente di aprire a questo punto una digressione circa l'esatto assetto degli antichi teatri di Corte. Il rilievo si riferisce presumibilmente ad una prima fase progettuale, allorché, non ancora demolita la porzione di edificio contenente la cavea, si pensò di ridurre a magazzino la sola zona del palcoscenico. Ebbene, nel preventivo di spesa "per la demolizione del Teatro Regio nelle parti comprese fra i muri di perimetro", pure datato 6 ottobre 1876, si parla di due palcoscenici, proponendo la demolizione delle murature "che dividono il palcoscenico alto da quello basso". Dunque, l'ipotesi della compresenza di due diversi edifici teatrali nello stesso perimetro murario sembra prendere consistenza. Il rilievo dell'Andreani mostra infatti come, alla zona del palcoscenico (basso) del teatro piermariniano potesse essere giustapposta, previa demolizione del diaframma murario, la zona del palcoscenico alto, ovvero una struttura preesistente collocata al primo piano.

A questo punto anche un particolare del disegno eseguito nel 1855 da Leopoldo Furlani acquista significato: la scala rappresentata al centro della parete di fondo del palcoscenico piermariniano, ascendendo di circa 4 metri, immetteva in un secondo vano palcoscenico. Alla luce di queste osservazioni diventa significativa, retrospettivamente, anche la già notata profondità del palcoscenico del teatro del Bibiena.

Ma la conferma di quanto finora congetturato arriva dalla lettura di alcuni inventari del teatro redatti tra Settecento e Ottocento a corredo degli atti di concessione in appalto dell'edificio agli impresari teatrali. Il più antico fra quelli rinvenuti reca la data del 7 novembre 1789 e, a soli 6 anni dall'inaugurazione, ci offre una puntuale descrizione del teatro del Piermarini. Infatti, benché scritto in un referenziale stile notarile, il documento ci regala l'affascinante sensazione, quasi in un moderno esperimento di realtà virtuale, di percorrere in ogni sua parte l'edificio, tanto è precisa l'indicazione delle coordinate spaziali, e tanto è meticolosa la descrizione di ambienti, arredi, attrezzature teatrali e suppellettili dell'epoca. Ebbene, apprendiamo così che per accedere alla sala retrostante il palcoscenico "evvi una porta grande di piella connessa a traversi" e che "s'apre questa ordinatamente mediante una portella in due ante formata nel mezzo. Ma in circostanza di voler rendere maggior sfondo al palco s'apre tutta, cioè due ante grandi s'appiantano ai muri della sala lateralmente e la scassa superiore fatta a ribalza va raccomandata ai pajoli". Dunque l'ampio spazio retrostante il palco, che aveva a suo tempo ospitato palcoscenico e cavea dei teatri all'antica del Bertani e dell'Andreasi, veniva fruito occasionalmente come profondissimo cannocchiale prospettico della piermariniana. Allorché invece le esigenze sceniche non lo richiedevano, la cosiddetta "sala retro il palcoscenico" era fruita come atelier ad uso dei pittori scenografi.

Apparatuses 167

Sempre l'inventario del 1789 offre un'altra preziosa notizia, a sostegno, se ve ne fosse ancora bisogno, dell'ipotesi dei due teatri giustapposti "rimpetto al palco scenario ed alla porta grande d'entrata nella sala (retrostante il palcoscenico) ... (vi è) una porta grande che dà uscita retro il Teatro, a comodo ingresso e regresso de' virtuosi ed inservienti, e che mette capo ad un scalone di marmo". Quest'ultimo manufatto, ben visibile e nel rilievo di Giuseppe Bianchi del 1775, nel disegno allegato al rogito della compravendita del 1896 e nel rilievo dell'Andreani del 1897, era dunque lo scalone che, ascendendo al primo piano adduceva nel teatro all'antica del Bertani i Gonzaga e i nobili cittadini, provenienti gli uni dal Castello e dal porticato della piazza, gli altri dal voltone vianesco. Diventa dunque illuminante, a questo proposito, l'affermazione di Ercolano Marani, [...], a proposito del monumentale portale d'ingresso a piazza Castello progettato dal Viani: "ebbe precipua funzione di architettura d'invito a un nuovo teatro".

Ma torniamo ora alle vicende delle demolizioni.

Preso atto della necessità, evidenziata nella relazione dell'Andreani di procedere allo smantellamento della struttura teatrale, i lavori per le demolizioni delle parti cadenti del Teatro Regio furono affidati alla ditta di Carlo Pavesi, aggiudicataria di un primo appalto. Il capitolato previde, per una spesa complessiva di 5.200 lire, la demolizione del tetto, del cornicione, delle murature del perimetro per l'altezza di un metro e di tutte le parti ammalorate. Il medesimo appalto affidava alla ditta anche la ricostruzione del tetto con materiale di spoglio e l'apertura di due porte di comunicazione con via S. Giorgio e piazza Castello, ma tale parte delle opere non risulta citata, e dunque nemmeno realizzata, nel verbale di collaudo definitivo del 4 ottobre 1898.

Le demolizioni furono eseguite tra il gennaio ed il marzo 1898: i materiali di risulta furono in gran parte recuperati ed accatastati per il riuso, mentre i calcinacci, liberati dalle incannicciature, furono sparsi nella platea. La notizia della decretata scomparsa del teatro fu chiosata da un importante articolo, probabilmente di Stefano Davari, dal titolo "Intorno alle fabbriche dei Teatri Ducali di Mantova", pubblicato sulla "Gazzetta di Mantova" del 15/16 gennaio 1898.

Circa la destinazione degli arredi, rimane solo la testimonianza della vendita del palco reale, progettato da Paolo Pozzo ed ornato di stucchi dorati e dipinti raffaelleschi di scuola mantovana del finire del '700, all'antiquario di Venezia Leone Polacco.

Unici eccezionali documenti visivi del teatro piermariniano restano due fotografie, scattate a cura della stessa ditta Pavesi prima di procedere alle demolizioni, e pubblicate sulla "Gazzetta di Mantova" del 30 marzo 1969. Mentre eseguivano le demolizioni, si avviò la discussione intorno all'assetto definitivo da dare al sito e alla destinazione dello stabile. Dalla documentazione conservata nell'Archivio Storico del Comune si ricava che l'Ufficio Tecnico aveva presentato all'attenzione della Giunta e del Consiglio Comunale tre distinti progetti: il primo ed il secondo, entrambi per una spesa complessiva di 15.300 lire, prevedevano, previa ulteriore demolizione della parte dello stabile contenente la platea e fronteggiante il Castello, la suddivisione della restante area di circa 972 metri quadrati in tre piani mediante due solai. Unica differenza tra i due progetti, la foratura dei muri perimetrali: con alternanza di porte e finestre al piano terreno il primo, con sole arcate il secondo. Il terzo progetto prevedeva invece, per un importo di 10.000 lire, la conservazione di tutto il perimetro della fabbrica (m 75x18) per un totale di 1350 metri quadrati e la suddivisione dello spazio in soli due piani con un solaio poggiante su 26 pilastri in muratura.

Il dibattito in corso, caratterizzato dalle relazioni del consigliere Poma e dall'assessore Lanza, rispettivamente favorevoli alle due soluzioni prospettate dai tre progetti, quella che potremmo definire "breve" (primo e secondo progetto) e quella "lunga" (il terzo progetto, propugnato dalla Giunta), si snodò nelle sedute del Consiglio Comunale del 30 marzo e del 12 e 19 aprile 1898.

La scelta cadde sulla soluzione "breve", in particolare sul secondo progetto si deliberò infatti di demolire la parte del fabbricato antistante il Castello, di destinare il piano terreno a mercato dei bozzoli, di costruire un solo piano rialzato ad uso magazzino di materiali del Comune, di erigere una nuova facciata verso campagna con grandi aperture, e di creare grandi arcate in asse con gli intercolumni del portico di piazza Castello: destinare il tutto per una spesa complessiva di 15.000 lire.

L'edificio sarebbe venuto così a configurarsi, al piano terreno, come grande loggiato ad uso commerciale a tre navate, scandito da due file di nove pilastri ciascuna: una sorta di prolungamento, al coperto, dello spazio aperto di piazza Castello tradizionalmente destinato a fiera.

(b)   La costruzione e le trasformazioni

La nuova tranche di lavori, consistenti nella continuazione delle demolizioni per lo "scoprimento del Castello dei Gonzaga" e nella costruzione del Mercato dei Bozzoli, ebbe inizio nel maggio 1898, sempre a cura della ditta Pavesi aggiudicataria di un secondo appalto.

Il progetto di massima da realizzarsi è sicuramente identificabile con il disegno datato 12 aprile 1898 a firma di Carlo Andreani, a sua volta collegato al "Capitolato Speciale" del 30 aprile 1898. Tale capitolato ebbe per oggetto "la formazione di un mercato coperto in aggiunta all'area di quello dei bozzoli nella vicina piazza Mercato, con sovrastante locale ad uso magazzino comunale". L'articolo 8 "Descrizione dei lavori" previde, tra l'altro, la "formazione di aperture di porte al piano terreno sulle tre fronti scoperte": in tal modo il fabbricato, compiuto entro il novembre dello stesso anno, sarebbe venuto a configurarsi come grande loggia commerciale.

Ma da una relazione presentata dall'assessore Camillo Colorni nella seduta di Consiglio del 17 novembre 1898 apprendiamo che, in corso d'opera, s'erano resi necessari alcuni interventi non previsti, quali intonacature esterne e tinteggiature, nonché la posa di cancellate alle aperture verso campagna e su via S. Giorgio: nel corso della medesima seduta si deliberò inoltre il completamento del sistema di cancellate anche lungo piazza Castello. Queste opere, che di fatto modificarono in fieri il progettato spazio libero e passante del loggiato, furono dettate da ragioni "di estetica, di decenza e di moralità", rimanendo la nuova costruzione completamente indifesa ed in balia dei malintenzionati. Ci si avvide inoltre, fin da quel momento, di un inconveniente del grande fabbricato: l'eccessiva esposizione agli agenti atmosferici dei lati est e nord (verso campagna e via S. Giorgio), nonché la carenza di luce naturale.

Apparatuses

Alcuni degli interventi successivi mirarono proprio a correggere tali difetti. È infatti del novembre del 1900 una perizia di spesa per "la formazione di 4 finestre da metri 1,60 x 3,90 inscritte nelle 4 grandi arcate verso campagna, ... ciò per togliere il lamentato inconveniente che le forti correnti d'aria smuovono gli ammassi dei bozzoli e igienicamente sono di danno alle donne occupate nella selezione dei bozzoli stessi. Non si propone l'otturamento delle altre aperture sostituendole con finestre perché il mercato allo stato attuale è deficiente di luce".

Continua così, tra formazione di aperture a rottura e tempestive tamponature, la tormentata storia delle poderosa mura perimetrali dell'edificio che niente impedisce di identificare con quelle originarie cinquecentesche, per quanto provate da incendi e trasformazioni, contenenti il primo teatro di Corte e l'armeria dei Gonzaga.

Nel corso del 1901 la stessa ditta Pavesi venne nuovamente incaricata di togliere le cancellate da poco installate e di ridurre a finestre una decina di aperture; in seguito, nel novembre dello stesso anno a causa di un episodio di vandalismo, si procedeva alla chiusura con finestre di altre cinque arcate. In tal modo, a quanto è dato di capire, l'edificio veniva completamente modificando il proprio aspetto, presentando ora finestre al posto delle arcate e due soli cancelli carrabili lungo i lati lunghi.

Giungeva frattanto a maturazione il progetto di prolungamento e completamento del mercato coperto verso piazza Legna, il lato ovest della fabbrica che ora affianca il voltone del Viani.

Nell'ottobre del 1901 l'Ufficio Tecnico presentò alla Giunta "il tipo per allungare il mercato coperto" con riserva di rassegnare successivamente "il progetto di dettaglio cogli studi pel prolungamento della facciata"; contestualmente si provvide ad inoltrare agli affittuari degli edifici esistenti nell'area tra il mercato ed il voltone l'ordine di liberare i locali entro il 1° maggio 1902.

Una seconda fase costruttiva ebbe inizio nell'autunno del 1902 allorché, con un terzo appalto, furono affidati alla ditta di Antonio Madella i lavori consistenti nella demolizione di alcuni edifici rimasti tra il voltone d'ingresso a piazza Castello e piazza Legna, nel prolungamento del fabbricato sull'area così sgombrata e nella costruzione dell'attuale facciata.

La lettura del "Capitolato speciale per l'appalto dei lavori di completamento del Mercato Bozzoli in Mantova", per quanto redatto in linguaggio tecnico edilizio, fornisce interessanti notizie per la conoscenza delle caratteristiche del nuovo tronco di edificio. Le fondazioni si sarebbero elevate da uno scavo della profondità di 5 metri; cornici, fasce, stipiti e bugnati avrebbero dovuto essere identici a quelli esistenti sul prospetto verso via S. Giorgio; in marmo dì Verona si sarebbe costruita una gradinata grande a terreno e la scala che metteva al piano superiore; si prevedeva inoltre una spesa maggiore di mano d'opera per la formazione di una finestra trifora con sagomatura e colonne; era contemplata infine la fornitura di uno stemma del Comune di Mantova in cemento armato, conforme al disegno fornito dall'Ufficio Tecnico, da issare sulla facciata.

Il disegno "allegato A" cui si allude all'art. 19 del Capitolato rappresenta la facciata ed il portale d'ingresso del mercato, con scritta e stemma cittadino: il prospetto fu realizzato con modifiche, in particolare alle cornici mistilinee delle finestre rettangolari al piano terra. Ma il testo del Capitolato desta interesse anche per la descrizione delle parti da demolire, tra cui "una scala di marmo formata da 30 gradini su volta rampante". Demolizione che il computo metrico relativo alle opere eseguito dell'impresa Madella fino al 1° dicembre 1902, precisa come "demolizione scalone".

Sono queste le ultime testimonianze del manufatto che, come constatato nell'esame degli inventari sette-ottocenteschi del teatro di Corte, adduceva alla cavea del teatro all'antica di Bertani.

I lavori per il prolungamento ed il completamento dei Mercato dei Bozzoli, interrotti durante i mesi invernali, furono compiuti nella primavera del 1903, e collaudati definitivamente il 27 maggio; ma solo nell'aprile dell'anno successivo l'edificio poté dirsi finalmente completato con la messa in opera, presso l'ingresso ad arco monumentale in piazza Legna, di un grande cancello in ferro di forgia liberty, realizzato da Carlo Battistoni.

Negli anni successivi l'edifico commerciale fu oggetto di semplici interventi manutentivi fino al 1922, allorché, constatandosi la progressiva sottoutilizzazione del salone, tenendosi il mercato dei bozzoli un solo mese all'anno e per di più sotto i portici di piazza Castello, l'ingegnere comunale Aldo Badalotti rassegnò un progetto per la riduzione della campata su via S. Giorgio a botteghe. Esse, previste nel numero di otto, si affiancarono alle tre già esistenti in prossimità di piazza Legna (segnate b nel progetto). A disposizione del mercato coperto sarebbero pertanto rimaste altre due campate che ricevevano aria e luce dai finestroni su piazza Castello.

I nuovi lavori non dovettero intaccare l'aspetto esterno a bugne del prospetto: comportarono però il rialzo del pavimento delle botteghe al livello del piano stradale, essendo questo notevolmente più alto del piano di posa del salone, e la demolizione delle "murature entro gli archi di progetto della via S. Giorgio".

Dalla documentazione amministrativa relativa allo stabile conservata presso l'Archivio Storico Comunale non risultano, da questa data fino alla metà degli anni '60, altri interventi edilizi tali da modificare forma o assetto del fabbricato così come ricostruiti fino ad ora. Accanto a lavori di ordinaria e straordinaria manutenzione (restauri del tetto, tombinature, rifacimento pavimenti, revisione impianti elettrici, costruzione servizi igienici ecc.) sono documentate opere di riparazione e adattamento del cancello d'ingresso in ferro, a causa di danni provocati tra il 1944 e il '47 da bombardamenti di veicoli pesanti.

L'edifico esercitò la sua funzione di mercato dei bozzoli fino al 1930. Finché, con deliberazione del 30 dicembre, il podestà Parmiggiani, trasferendo il commercio nei locali del mercato bestiame del Te, destinò lo stabile a mercato ortofrutticolo, funzione che molti Mantovani ancora oggi ricordano.

Documents

Letter from Mantua dated 27 October 1532. The writer is Federico II Gonzaga, Duke of Mantua, while the addressee is his mother Isabella d'Este. The answer of the Marquise, dated 28 October 1532, is transcribed in the Appendix.

Apparatuses

ASMN (Archivio dí Stato di Mantova), A.G. (Archivio Gonzaga), b. 2132, cc. 315v-315v.

Illustrissima et cetera.

Havendo a venire qui la Maestà Cesarea, come Vostra Excellentia deve sapere, et essendo io per volere honorare quella in tutto quello che serà possibile, sono stato advertito dallo illustrissimo signor don Ferrando mio fratello che una delle cose più accete che passi fare sie qualche comedia, et così ho determinato. Et perch'io non trovo loco ove farla recitare, che sia più al proposito di quella sala di Vostra Excellentia, che è di capo le stantie delle sue damigelle, ho fatto expedire volando questo cavallaro per pregare quella che quando non li sia molto incomodo, così de la stantia come de li strepiti nel farli lavorare, che la sia contenta compiacermine, advertendola anche che quando la sia contenta di accomodarmine, bisognarà che anche la mi acomodi, per quelle sere che si recitarano le comedie, dì quella camara ch'è di capo la sala, perché l'impedimenti de la scena non lassarano che si passi entrare per il poggio, per il che serà necessario entrare per la ditta camera. Sì che Vostra Excellentia si dignarà di subito rimandar indrieto questo cavallaro, con quella expeditione che li parerà, perché, se ben so che quella sia per essere qui presto, non di meno l'è tanta l'angustia del tempo che ho voluto guadagnar questo puocho, et in bona gratia de Vostra Excellentia et cetera.

Da Mantova, agli XXVII di octobre MDXXXII.

Letter from Mantua, dated 3 November 1532. Ippolito Calandra writes to Federico II Gonzaga, Duke of Mantua.

ASMN, A.G., b. 2517, cc. 63r.-66r.

[…] Questa matina [ho i]nteso come vostra excellentia serà qui mercori che viene. [Per a]visare Vostra Excellentia di tutto, quanto sia al castello [è] aconcio et adatato benissimo in ogni loco, che cosa alcuna non vi manca nelle stantie nove, li corami novi à aparato quatto stantie, le più grande, et stanno molto bene con le sue vedriati su le finestre; l'altro camarino io l'ho fatto a parare di alcuni pezi di tapezarie che si confà con la spalera bella ch'è in casa, dove fu aparato l'altra volta che sua maestà fu a Mantua. Il camarino dil Poggio, quale sta molto bene, poi li ho fatto metere quelle tavole de tarsia che fece maestro Richo Todesco, che stanno bene; quello altro camarino, che ha fatto fare la signora duchessa, hogi non è anchora fornito, ma dimane se finirà et anche lo farò aparare lui di qualche cosa con le store sotto per la umidità di muro fresco. Circa a l'arteliaria tutta è aconcia benissimo lì, in quel loco che ordinò vostra signoria, dove sta melio che non faceva l'altra volta: prima tutta la grossa è tirata alla filla e poi, dinanti, vi sono li falconi et tiene di spacio, cominciando lì di sotto al loco, dove sta dicta arteliaria, fina [sotto] al muro dil giardino di Nardo dove so[no] et viene ad essere dui ordine, uno [di pesante], l'altro di sutila, in fogia di luna. […] Una dificultà me pare a me li resta, sì come [il mio] poco iudicio et anchor d'altri: vostra signoria ha da [saper] che quanto si recita

una comedia, bisogna che cussì [quelli] che sono dentro la sena e che governano la come[dia] oldanno cussì bene come quelli sono di fori che l'as[colta]no et più toca a loro che ad altri per sapere quan[do] hanno da mandare fori et dentro a posta sua [...] che una grande importanza, l'altra sì come li re[ci]tatori vanno recitando di dentro, continuamente lo va[no] seguitando uno o dui con il libro in mane et. [se], per dísgratia, il fala o che se scordi, come sempre acade in simile caso, subito quello di dentro lo dice et li racorda et aiuta quello hanno da dire che, quanto [non] se facesse cussì, non seria persona alcuna che non falasse o che non se perdesse. Per remediare a questo, si fa le [c]ase di asso, con le sue finestre et strate et di sopra aperto, dove facilmente quelli de dentro odono il tutto, sì come quelli fori et se sanno governare in tutte le cose sue et mandarli fori a tempo, sì come è il bisogno, et maxime questa comedia Calandra che li interviene le più dificile cose che altra comedia, sì nel dire, come ne li vestimenti, che vi sono almeno cinque che bisogna vestirli et disvestirli de diversi habiti in più volte, dove vi bisogna una pronteza et una presteza miravile et sopra il tutto buonissima orchia di quelli che la governano, et maxime nel vestire, quale so tocar a me, vi bisogna grandissimo fastidio et bono ochio et più le orechie.

Hora la dificultà è questa: messer Iulio Romano ha mutato tutto quello proposito et ordine con che se risolsse di fare quanto Vostra Excellentia se partite da Mantoa; primo, lui ha piliato l'altra parte di la stalla a mane stanca, quale sta bene per il strepito che se haveria sentito dinanti, et fa la sena da recitare, la fa al longo, a man sinistra vers[o il por]tico et vole dipingere le case sul muro et [che li] recitatori habiano da stare in d'el portico [...] usire fori per il muro sopra uno palco di quat[ro teste] largo, dove haverano da recitare, dove se, per sorte, quelli o quello che serà fori a recitare et che ve[...] a falare o scordassi qualche cosa non serà già mai [possi]bile che quelli che governeranno possa racordarli né met[...], se si falarano, per la grossezza del muro, quale è di qu[atro] teste grosso, che non se sentiria, se bene se cridass[ero] mai si potesse, che, se per disgratia acaschasse questo, se ritrovaria molto impaciato et rimedio non [...] seria di aiutarli, perché mai non se saperà queli che la go[verna]no quello li averano da fare, che certo che non se oldirano; che, con tutto quanto le case sono di asse, si a f[atica] a oldire, pensa mo' Vostra Excellentía quello farà hora, havendolí per tramezo uno muro di quatto teste. Che so dire a vostra signoria, che mi dubito certo che haveremo poco honore et, per dire, melio seria a farla di asse, che farla più bello vedere con le sue case di relevo, como altre volte è state fatte, che fare questo, et so certo che seria manco [...] a e più presto si farà et se torà via tutta la [diffi]cultà che li viene ad essere, che questo dico certo a vostra signoria che tutti se ritrovaremo impaciati. Il tutto [ò] dítto a esso messer Iulio Romano et, ditole tutte [le] ragione, non n'è possibile che la volia intendere, vole fare a suo modo, e facia: solum me rincrese che vostra signoria non buta via la spesa et che poi non si resta vergognato; in questo lui dice che li farà dí molti busi per il muro con sopra di la carta dipinta che se poterà oldire.. Io dico questo a vostra signoria, se ben vi fosse trenta milia busi, che mai non se oldirá quello dirano, cioè quelli che la governano, essendo questo muro dinanzi di quatro teste. Io ho voluto il tutto di questa celata avisare Vostra Excellentia a ciò che la

Apparatuses 173

sapia il tutto et che la non possa dire se lo avesse saputo, non se seria fatto. Io credo bene, quanto sia per le fabriche et disegni, messer Iulio se intenda melio di me, ma, quanto sia per governare la comedia et dire quello li bisogna, se intenda poco alle ragione ch'io le ho ditte, che lui non le vole capire.

Letter dated 12 February 1501, transcribed in D'Ancona, op. cii., II, pp. 382–83.

[…] Doj bande era scena data ad actorij et recitatorj: le doe altre erano ad scalini, deputati per le donne et daltro, per todeschi, trombecti et musici. Al jongere del'angulo de un de' grandi et minori lati, se vedevano quactro altissime colonne colle basi orbiculate, le quali sostentavano quactro venti principali: fra loro era una grocta, benché facta ad arte, tamen naturalissima: sopra [la quale] era un ciel grande fulgentissimo de varij lumi, in modo de lucidissime stelle, con una artificiata rota de segni, al moto de' quali girava mo il sole, mo la luna nelle case proprie: dentro era la rota de Fortuna con soi tempi: regno, regnavi, regnabo: in mezzo recideva la dea aurea con sceptro con un delphin.

Ferrara, 18 March 1532, Ludovico Ariosto to Giovan Jacomo Calandra.

ASMN, Autografi, busta 8, c. 458.

Mag[nifi]co M[esse]r Gíovan Jacomo Calandra mio […] io mando per lappostator de la l[ette]ra di V[ost]ra Ex[cellen]tia quattro commedie cioè tutte quelle che mi trovo mai haver fatte. Quella sarà contenta di donarle da mia parte all'Ill[ustrissi]mo Signor Duca. Sio ne finirò un'altra et già molto cominciai e menatala et pezzo inanzi per altre occupazioni la missi da parte; io ne farò coppia a Sua Ex[cellen]tia; adesso io sono così occupato per mettere un'altravolta il mio furioso a stampa con alquanto di additione et non posso attender ad altro. E sin queste commedie troverete qualche errore circa l'osservatione della lingua scusatemi che s'anchora chio gli habbia veduti non ho havuto tempo di correggerli oltre quello chio scrivo al S[igno]re Duca. V[ost]ra S[ignor]ia lo pregarà da mia parte che per in avisanza di chi havran le commedie ne le mani non si lascino si che vadano a stampa come sono andate talune con mio grande dispiacere et à V[ost]ra S[ignor]ia mi offro e raccomando.

Ferrara XVIII Martii MDXXXII.
   Di V[ost]ra S[ignor]ia.
   Ludovico Ariosto.

Ferrara, 18 March 1532, Ludovico Ariosto to Federico II Gonzaga, Duke of Mantua.

ASMN, Autografi, busta 8, c. 459.

Illustrissimo et Ex[ellentissi]mo Signor mio io mando à V[ost]ra Ex[cellen]tia per questo suo gentilhuomo […] tutte le commedie che mi trovo haver fatte: che sono quattro come io promissi di far per una mia chio scrissi a Braghino: et hora da

174 Apparatuses

messer Giovan Jacomo Calandra mi sono state da parte di V[ost]ra Ex[cellen]tia domandate. Due ci sono e non credo che quella habbia più vedute: latre anchora che siano a stampa per colpa di persone che me le rubano non sono però nel modo in che io li ho ridotte: massimamente la cassaria che tutta è quasi rinnovata. Selle satisfaranno à Vostra Ex[cellen]tia rihaverò piacer grandissimo. Quella suplico che sia contenta di non lasciarle andar in modo che sieno stampate un'altra volta [...] et non credo che le stampassimo più corrette ch'abbian fatto laltravolte, io ricognosco dentro detti errori circa la lingua che per trovarmi hora occupato et altro, non ho havuto tempo di correggerli, et ancho che li ha malscritte non ci ha usato quella diligentia, ch'auria possuto; et io per questo huomo di Vostra Ex[cellen]tia non me ne venga senza non ho tempo di rivederla altrimenti et piuttosto voglio ch'ella le habbia hora non così ben scritte, che indugiando darle sospetto chio sin men pronto allo servitio suo di quello che è mio debito disservir in buona gratia alla quale mi do et raccomando sempre.

Ferrara XIII Martii MDXXXII.
    Di V[ost]ra Ex[cellen]tia.
    Devotissimo Servitor Ludovico Ariosto.

Ferrara, 5 April 1532, Ludovico Ariosto to Federico II Gonzaga, Duke of Mantua.

ASMN, Autografi, busta 8, c. 460.

Illustrissimo et Ex[ellentissi]mo Signor mio Osser[vissi]mo mi duole che le mie commedie per esser in versi non habbiano satísfatto à V[ost]ra Ex[cellen]tia; me pareva che stassi così meglio che in prosa, ma li giudicii son diversi. Le due ultime io le feci da principio nel modo che stanno e duole di non averle ancho fatto in prosa per haver potuto satisfare a quella, laquale si acontenta di accettare il buon animo. Io le riferisco gratie che le habbia. poiché non fanno per lei, rimandate subito. Di buona gratia dela quale mi raccomando sempre.

Ferrara V Aprile MDXXXIII.
    Di V [ost] ra Ex [cellen]tia.
    Devotissimo Servitor Ludovico Ariosto.

Montagnana, 28 October 1532, Isabella d'Este, Marquise of Mantua, to Federico II Gonzaga, Duke of Mantua.

ASMN, A.G., b. 2132, cc. 326–27.

Ill[ustrissi] mo et Ecc[ellentissi]mo S[igno]r figliuolo mio hon[orevole].

La l[ette]ra di Vostra Ecc[ellen]za mandata à posta con la quale mi ricerca la sala et stantia mia per bisogno à una comidia ch'ella disegna di far in questa venuta de la M[aes]tà [...] m'ha ritrovata qui in Montagnana donde sun partir fra due per andar questa sera ad alloggiar à Sanguanito et dimane arrivare con l'aiuto di Dio,

Apparatuses

à Mantoua. Io non vi farò perhora altra risposta riserbandomi di satisfar à prima a buon conviti à la quale fra tanto mi raccom[an]do. Questo posso.

Di Montagnana alli XXVIII d'ottobre MDXXXII.
  Amorevoliss[imamente] et buona M[ad]re.
  La Marchesa di Mantoua.

Mantua, 29 October 1532, Federico II Gonzaga, Duke of Mantua, to Titian.

ASMN, A.G., Copialettere, b. 2934, lib. 305, c. 191r.

Messer Ticiaro, amico et cetera. Vi priego che mi mandate qui quel pittore piacevole che sapeti, per fare qualche bel spettaculo alla maestà cesarea, in alcune comedie che ho dissegnato di fare alla venuta di quella, che mi farete piacere grande, et alli commodi et cetera.

Mantuae, XXIX octobris MDXXXII.

Mantua, 1 November 1532, Ippolito Calandra to Federico II Gonzaga, Duke of Mantua. ASMN, A.G., b. 2517, c. 136r.

Illustrissimo Signor mio Signor Singularissimo. Il castello a questa hora è quasi aparato dove fu belissimo vedere quelle tre camare una dreto l'altra aparata di quelli veluti verdi et tela d'oro. Il camarino dal Sole è aparato, sì come l'altra volta, di damasco cremesino e veluto tané con quelli dentelli di tela d'oro; il camarino dal Pozo li ho fatto metere quelli aparamenti, che portò la signora duchessa, da alto a basso di tela d'oro et aregento, quali stanno benissimo, dove penso che sua maestà più starà in quello loco che in altro et maxime hora che si pò fare fogo. Rora siamo dreto ad aparare in la fabrica nova con li suoi corami, dove mi penso che starà bene; li dui camarini da basso, per andare a l'oratorio, sono aparati uno de quelli aparamenti intaliati di brocato et veluto cremesino, quali erano in d'el camarino di la signora duchessa, l'altro è aparato di veluto verde et tela d'oro di quelli sono avanciati alle camare, quali stanno bene. In sul Te è aconcio ogni cosa et le due camare anche bisognando per il signor Ferrando. La Bianchina et Batista tutti se sono retirati di sopra in una stantia et li resta di sopra cinque camare vote. Tutta l'artelaria è fori et aconcia che sta bene et fra le roche di Santo Zorzo, che sono due, vi sono cento e venti code di ferro. Circa a l'aparato della comedia, messer Iulio non li manca di fare lavorare. Al tutto non si mancarà, baso le mane a Vostra Excellentia.

Mantue, primo novembris 1532. De Vostra Excellentia fidelissimo Hippolito Calandra.

Mantua, 20 December 1532. Payment warrant for scenographic equipment.

ASMN, Autografi, b. 7, c. 4 19r.

Magnifico domino texaurero generalle de l'illustrissimo signor duca nostro facia pagamento a maestro Vicenso Bersano depintor che àno depinto una tela larga e alta como era l'aparato de la comedia, depinta come paiesi e cavali e uno imperator a cavallo e cavali colegati e turchi colegati, d'acordo in schudi seti d'or in or como el spectabile messer Iulio Romano, superiore generalle de li fabriche a dì 15 de novembre 1532, monta libre 36 soldi 15 lulio Romano.

Franciscus Bruschus, notarius fabricarum, vigore bulete Zan Tartalia suprastantis, signate per spectabilem dominum Iulium Romanum, superiorem generallem fabricarum, sub die 20 decembris 1532, in lioro Regula, carta 24.

Fíat mandatum. Mandato illustrissimi principis et excellentissimi domini nostri domini ducis Mantue et cetera, magnificus tesaurarius generalis dare debeat magistro Vicentio Bersano libras triginta sex et solidos quindecim, prout supra patet, et ponat expensis.

Registratum. Franciscus Columba notarius scripsit XXIII decembris1532.

Petrus Gabloneta. Nazarius Scopulus. Nicolaus Aliprandus.

Posto a spexa al libro Riga, a carta 50.

Scarpa C (1965) Relazione di progetto del Teatro Carlo Felice di Genova.

Il compito che mi è stato affidato, com'è noto, è il restauro del teatro del Barabino. Penso che nello scegliere il mio nome si fossero implicitamente accettati i criteri del mio metodo di intervento in questi casi, che sono quelli del restauro critico. Sono criteri di rigoroso rispetto dei resti autentici e originali del monumento, e di libero inserimento delle aggiunte, integrazioni, suture, completamenti necessari a ridare funzionalità al monumento, in uno spirito di interpretazione dell'antico e di composizione unitaria, che è appunto il "momento critico" del mio intervento. Nel nostro caso ho inteso accentuare il carattere neoclassico e la funzione urbanistica del grande vaso—grosso modo cubico—rimasto ancora in piedi, inserendovi le tre parti del nostro teatro (ridotto, cavea, palcoscenico) in una articolazione che rispetta e direi accentua, sotto certi aspetti, gli assi strutturali dell'antico organismo, sfiora e non intacca le antiche murature, innova francamente funzioni e convenzioni secondo le esigenze di una società moderna, che si sviluppa libera nell'alveo storico, nell'ambiente che ha ereditato dal passato.

Le facciate delle parti nuove si saldano alle antiche senza mutare il carattere puristico della costruzione del Barabino, perché anche le nostre murature sono lisce e semplici, rivestite da intonaco appositamente studiato, solo ritmate da aperture nude e strette che sottolineano e rendono unitaria e compatta la saldatura del nuovo al vecchio.

Ho detto che in un certo senso ho voluto accentuare alcuni caratteri del monumento barabiniano; alludo alla mia cura di far corrispondere con più rigore i ritmi delle facciate esterne alle articolazioni della pianta, specie nell'atrio lungo, in asse con il pronao dell'entrata regia, che è stato ampliato, approfittando di una maggiore disponibilità di spazio rispetto alla situazione antica. E alludo anche alle accresciute funzioni urbanistiche che ho dato alle due entrate del teatro, collegate opportunamente con l'asse viario della galleria Mazzini, mediante una soluzione che amplia e vitalizza il traffico pedonale immettendolo nel monumento: il quale così da un lato non è più diaframma, bensì tramite fra la galleria, via Roma e la piazza De Ferrari e,

Apparatuses 177

dall'altro lato, da questa nuova funzione, vede accresciuto spazio e respiro dei suoi atrii ed accessi.

Ho avuto cura altresì che l'accrescimento di area a nostra disposizione, secondo il PRG, non si risolvesse in un mutamento di rapporti volumetrici; il che ho ottenuto, spostando di circa due metri verso sinistra l'asse dell'atrio da via Roma, ma non alterando l'altezza del culmine dei tetti; sicché nella asimmetria, o meglio accanto alla duplicità di assi strutturali tra le due facciate, già adottata dal Barabino, ho inserita una complementare asimmetria, una traslazione e slittamento degli assi fondamentali che accentua il carattere libero, a incastri lievemente scorrenti della pianta.

Ciò contribuisce alle possibilità di libero sviluppo implicito nell'antico organismo. Appunto lungo l'asse longitudinale, penetrante dal pronao all'entrata regia, si salda l'organismo nuovo all'antico; in questo settore si alza, su piani sovrapposti e mantenendo il verticalismo dinamico ed aperto del vecchio foyer, un ponte articolato a rampe, terrazze, atrii, foyers corrispondenti ai vari gradi della sala teatrale che fa cerniera tra le facciate barabiniane e i nuovi inserimenti.

La distinzione delle singole parti del teatro fa parte del tema dominante del nostro intervento. Lo spazio dell'antica zona d'accesso viene rispettato nelle sue strutture basilari, immettendo nei singoli vani maggiori le nuove funzioni (il bar nella sala A; la biglietteria a piano terra e i foyers nei superiori ov'era il grande foyer, cioè nel vano B; una scala per il loggione nella scala C; un grande atrio a torre in D): il tutto unitamente articolato, in servizio e direi quasi a sostegno della sala teatrale vera e propria. Questa trova posto nella prima metà dello spazio residuo a nostra disposizione (l'altra è destinata al grande palcoscenico che ha il suo limite rigoroso nella "linea di fuoco", e che si sviluppa secondo le particolari esigenze scenografiche illustrate nella relazione dell'Ing. Zavelani Rossi), e in questo grande vano parallelepipedo si flette secondo un disegno libero e nuovo.

E' come una grande platea articolata in due bacini semicircolari, isolati nello spazio cubico, come due grandi valve cocleari, incernierate in un punto centrale, ove trova posto la cabina di regia, protette sui lati da una cortina che, come fascia sferica, modella in curva lo spazio dell'aula lungo i lati e lo accompagna verso il soffitto, mantenendo libero negli spicchi d'angolo la gittata dello sguardo verso l'alto. I due spicchi della platea, e il grande ferro di cavallo del loggione, si incurvano con un'articolazione lenta e continua, che permette di collocare ognuna delle 1800 poltrone in una uguale posizione visuale rispetto al palcoscenico.

Collegando idealmente l'occhio di ogni spettatore con un punto centrale del palcoscenico, i punti di partenza di ciascuna di queste linee si inseguono secondo tante catenarie convergenti verso l'orchestra; si ripetono così idealmente le condizioni di una grande cavea classica. La scelta di questa forma della sala intende rendere simili il più possibile le condizioni dello spettatore; e questa disposizione dei posti a sedere se da un lato risponde a una condizione sociale tipica di una grande città moderna, ripete dall'altro lato una soluzione già adottata per il più classico teatro moderno per spettatori "uguali" del suo tempo, qual era l'Accademia Olimpica Vicentina. All'abolizione dei palchi corrisponde la rivalorizzazione del loggione, all'accrescimento dei servizi e delle comodità la particolare cura avuta per rendere intercomunicanti e liberamente percorribili tutte le parti della sala e degli accessi in una continuità dei passaggi che permette una visione itinerante e quindi intimamente penetrante di tutta la macchina architettonica: questi accessi e passaggi, a balconi e scale penetrano nella struttura stessa della curva della platea, e ne costituiscono ad un tempo le nervature e i supporti che la sostengono nello spazio.

Una soluzione innovatrice ho voluto dare al proscenio cioè a quel punto di sutura tra palcoscenico e aula degli spettatori, che nei teatri barocchi e neoclassici ha la forma dell'arco scenico. In un organismo a stacchi e incastri—ove la unità architettonica risulta da accostamenti risaltanti e non per trapassi e continuità di svolgimenti lineari—mi è sembrato che la soluzione nuova del problema consistesse, in un certo senso, nella strutturazione di quelle ali delimitanti la scena, che gli antichi chiamavano versure, secondo le funzioni ad essa proprie.

Su queste ali inquadranti lo spazio scenico, si innestano perciò quei particolari palchi visuali che nei vecchi teatri si chiamavano barcacci, si collocano i supporti e gli schermi per i riflettori di scena, trovano il loro punto naturale di sbocco e di arrivo i percorsi perimetrali dell'aula nei suoi vari gradi e livelli.

Anche nell'orchestra è stata data una sistemazione, che conferisce ad essa un particolare tono e quel rilievo che devono avere, a mio parere, in un teatro d'opera, la camera sonora e il luogo d'azione dei concertisti e dei direttori delle musiche.

Lo spazio riservato all'orchestra è ampio in sé, ed ha alle spalle un ulteriore anello di protezione e di isolamento risultante dello sfociare ed emergere da sotto la platea del corridoio di ingresso, che proviene direttamente dall'entrata principale.

Il cielo del teatro è a gradoni di vario rilievo, che articolando le strutture portanti del tetto, hanno una particolare funzione acustica, e offriranno la possibilità di creare una illuminazione indiretta e diffusa, in funzione rigorosamente architettonica.

Ho infatti studiato questo teatro come un grande spazio architettonico, valido in se stesso, come monumento da visitare e percorrere a somiglianza degli antichi, e non come un semplice complesso di funzioni e attrezzature sceniche. Di qui la cura, che si renderà più chiaramente evidente nella fase esecutiva del lavoro, per la ricerca di materiali e di opere d'arte eseguite da artisti particolarmente significativi del nostro tempo, sì da dare ai vari ambienti del teatro uno spiccato carattere espressivo; e sarà cura della Direzione delle Belle Arti del Comune, io spero, di consigliarci e guidarci nella scelta di questa ideale galleria moderna immessa nell'organismo stesso di un teatro.

In una città come Genova un ambiente come il Carlo Felice ha da svolgere, a mio parere, una funzione culturale propria e specifica, e in questo senso esso va in ogni sua parte studiato.

Il consolidamento delle strutture del teatro è stato studiato in via preliminare e in stretto rapporto al nostro intervento dall'Ing. Prof. Luigi Croce il quale dà relazione a parte del suo prezioso lavoro.

A questo proposito credo utile qui ricordare che le nuove strutture hanno permesso di ricavare sotto il teatro, in comunicazione con i ricordati accessi pedonali, una seconda sala teatrale per settecento posti.

L'attrezzatura scenotecnica è stata progettata con eccezionale impegno e maestria dall'Ing. Zavelani Rossi, il quale—come dirà nella sua relazione—ha inteso con questo suo lavoro dotare il Carlo Felice dell'optimun sia dal punto di vista tecnico che delle possibilità spettacolari.

E anche in questo caso voglio sottolineare il fatto che le nuove attrezzature non hanno creato deformazioni delle proporzioni originarie del teatro; la linea di gronda della torre sovrastante il palcoscenico si è alzata di un metro e mezzo, ma si è ristretta sia in lunghezza che in larghezza, sicché ora ha uno slancio maggiore e più puro, sottolineato dalla compattezza delle rivestiture a squame quadrate di ardesia che tutte rivestono e decorano la torre.

Images

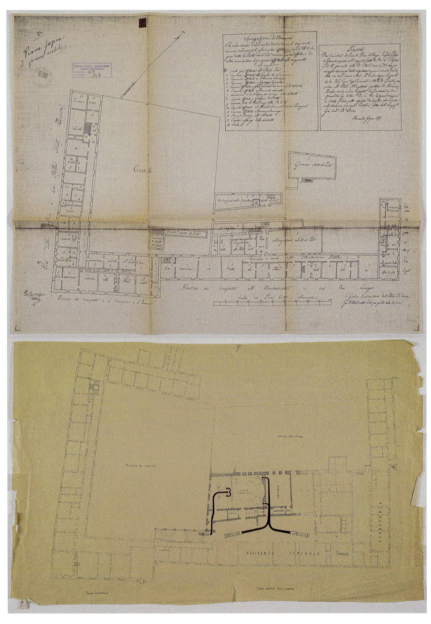

Planimetria ante e post operam della Sala del Consiglio, Palazzo Comunale di Ferrara [Archivio Storico Comunale di Ferrara, Cartografia, cart. 3/A, disegno n. 247 e 248]

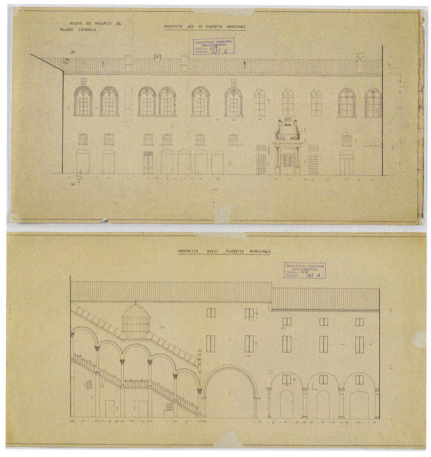

Prospetto esterno ed interno, Palazzo Comunale di Ferrara [Archivio Storico Comunale di Ferrara, Cartografia, cart. 3/A, disegno n. 281B e 300A]

Prospetto interno, Palazzo Comunale di Ferrara [Archivio Storico Comunale di Ferrara, Cartografia, cart. 3/A, disegno n. 281A]

Town Hall, Ferrara [elaboration by the author]

Apparatuses

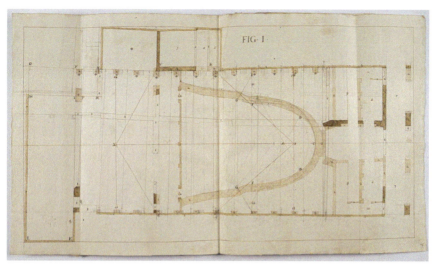

Pianta del livello di ingresso del Teatro Scroffa [Archivio Storico Comunale di Ferrara, Carteggio Amministrativo, secolo XIX, Teatri e Spettacoli, busta n. 71]

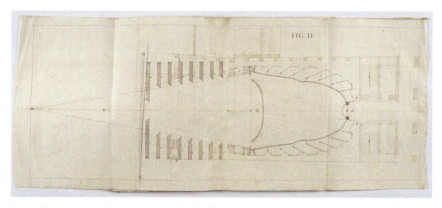

Pianta del primo livello del Teatro Scroffa [Archivio Storico Comunale di Ferrara, Carteggio Amministrativo, secolo XIX, Teatri e Spettacoli, busta n. 71]

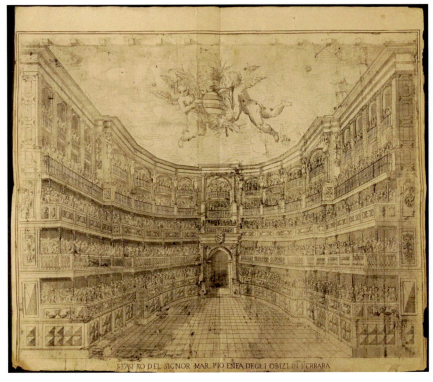

Disegno Teatro Obizzi (interno) [Biblioteca Comunale Ariostea, Fondo Antolini, 83]

Apparatuses

Antiche vestigia del Teatro Obizzi demolite in Maggio del 181. [Biblioteca Comunale Ariostea, Fondo Antolini, 83]

Disegno allegato al Capitolato speciale per l'appalto dei lavori di completamento del Mercato Bozzoli in Mantova, a firma Carlo Andreani, 28 agosto 1902, facciata [Archivio Storico Comunale di Mantova, V.3 1., busta n. 30]

Apparatuses 187

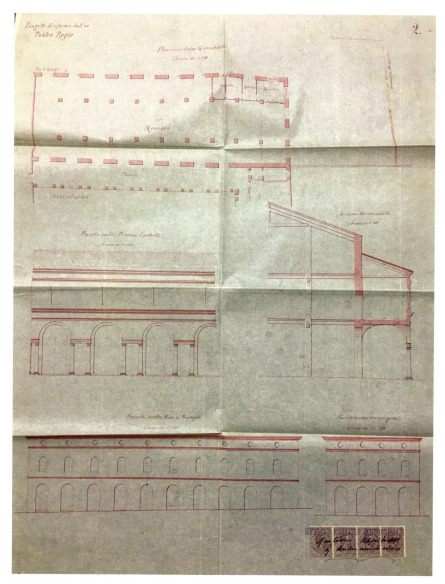

Disegno n. 2 allegato al Capitolato speciale per l'appalto dei lavori di completamento del Mercato Bozzoli in Mantova [Archivio Storico Comunale di Mantova, V.3 1., busta n. 30]

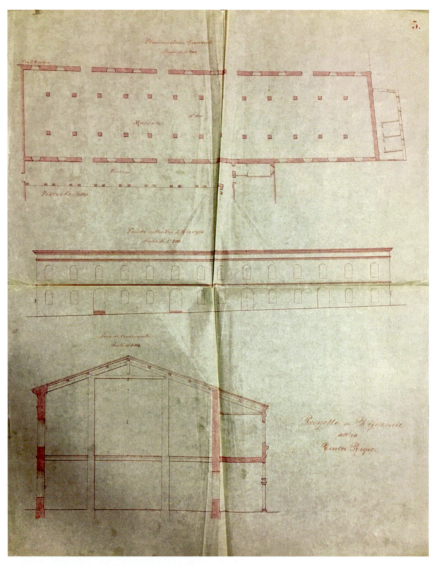

Disegno n. 5 allegato al Capitolato speciale per l'appalto dei lavori di completamento del Mercato Bozzoli in Mantova [Archivio Storico Comunale di Mantova, V.3 1., busta n. 30]

Apparatuses

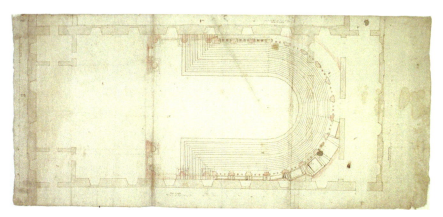

Giovanni Battista Aleotti, Pianta del Teatro del serenissimo Duca di Parma del salone grande fatto da ingegnero Argenta [1617–1619] [Biblioteca Comunale Ariostea, Raccolta Cartografica Aleotti, Classe I 763, tavola 162]

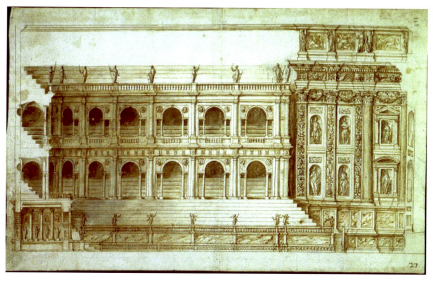

Giovanni Battista Aleotti, Alzato del Teatro Farnese di Parma, Progetto non realizzato [1615–1617] [Biblioteca Comunale Ariostea, Raccolta Aleotti, 165] Teatro Farnese di Parma [Biblioteca Comunale Ariostea, Raccolta Cartografica Aleotti, Classe I 763, tavola 165]

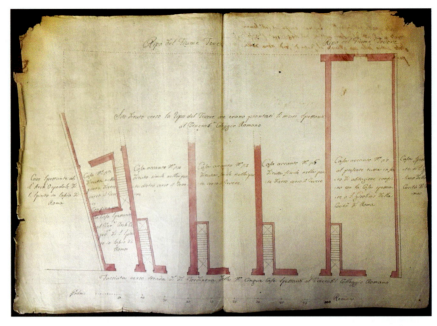

Pianta dell'isolato adiacente al teatro di Tordinona [Archivio di Stato di Roma, Camerale III (secoli XV–XIX), busta n. 2127]

Apparatuses 191

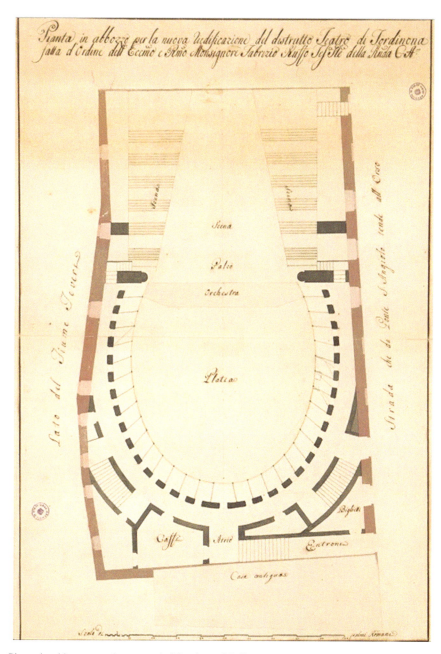

Pianta in abbozzo per la nuova riedificazione del distrutto Teatro di Tordinona fatta d'Ordine dell'Eccellentissimo, e Reverendissimo Monsignore Fabrizio Ruffo Tesoriere Generale della Reverenda Camera Apostolica [Archivio di Stato di Roma, Collezione I di disegni e mappe, cartella 89, foglio 626]

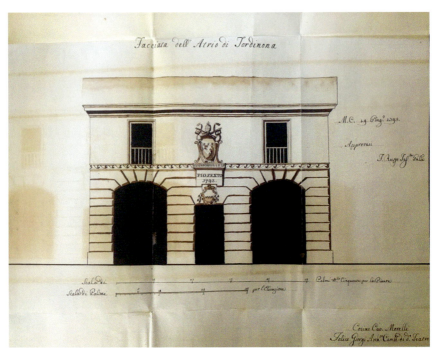

Facciata dell'Atrio di Tordinona [Archivio di Stato di Roma, Camerale III (secoli XV–XIX), busta n. 2127]

Apparatuses

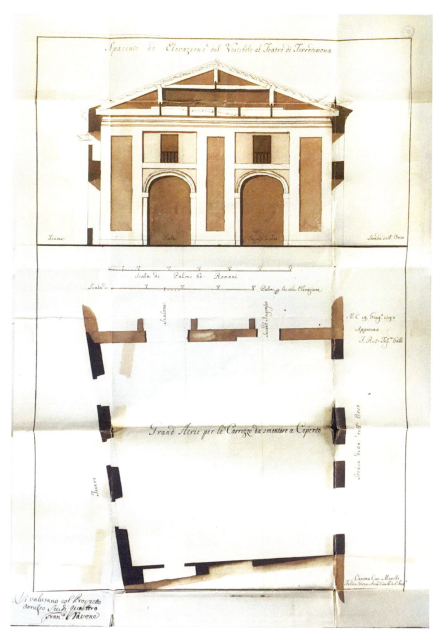

Spaccato ed Elevazione del Vestibolo al Teatro di Tordinona [Archivio di Stato di Roma, Camerale III (secoli XV–XIX), busta n. 2127]

Apparatuses

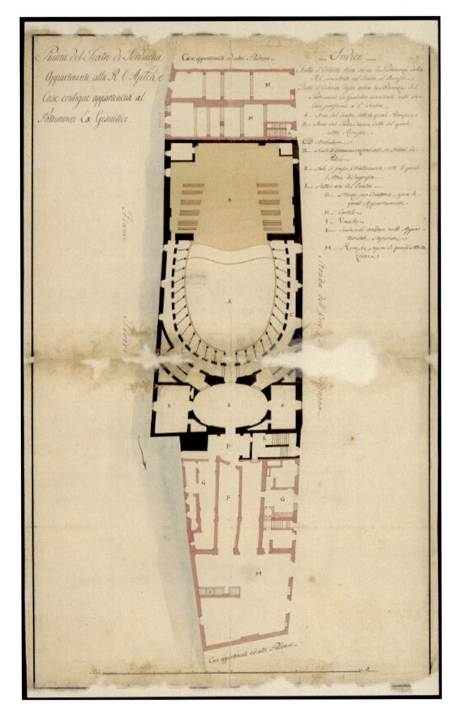

Apparatuses

Pianta del Teatro di Tor di Nona appartenente alla R.C. Apostolica e case contigue appartenenti al Patrimonio ex Gesuitico [Archivio di Stato di Roma, Collezione I di disegni e mappe, cartella 89, foglio 627]

Disegno tolto dalla Pianta di Ferrara del Bolzoni (1747). Quattro volte più grande dell'originale.
La parte colorita in rosso indica gli stabili che furono demoliti per la costruzione del Teatro.

Disegno della Pianta di Ferrara del Bolzoni (1747) con indicati gli stabili che furono demoliti per la costruzione del Teatro. [Biblioteca Comunale Ariostea, H.5.3 Vol.3 O.96]

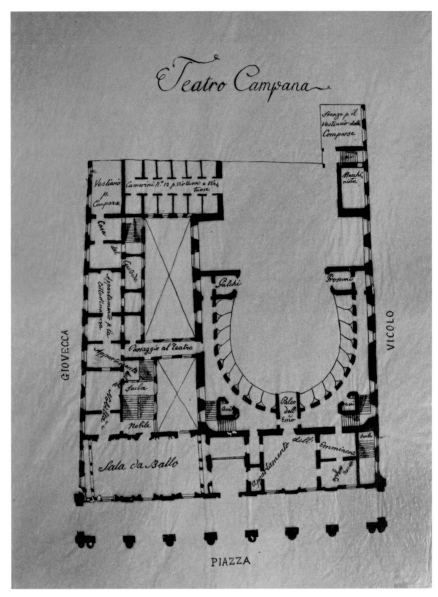

Primo Progetto del Teatro Campana (Comunale di Ferrara) [Biblioteca Comunale Ariostea, Il Teatro Comunale di Ferrara. Cento anni di storia (1798–1898). Memorie raccolte a cura di Giovanni Pasetti, Tullio Finotti, Luigi Villani, 1915. Parte prima, p. F, n. 18]

Apparatuses 197

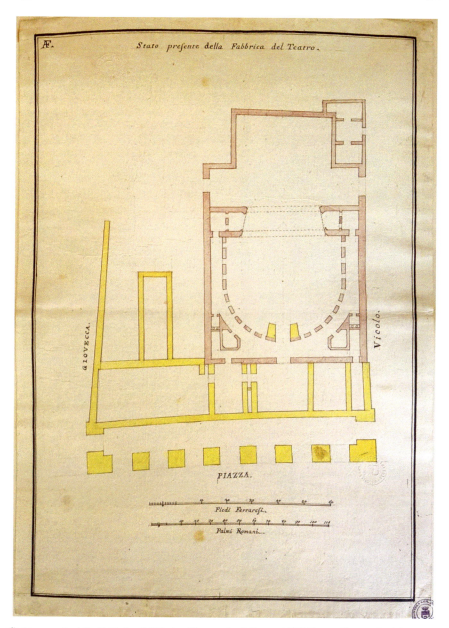

Stato presente della Fabbrica del Teatro (Comunale di Ferrara) [Biblioteca Comunale Ariostea, Il Teatro Comunale di Ferrara. Cento anni di storia (1798–1898). Memorie raccolte a cura di Giovanni Pasetti, Tullio Finotti, Luigi Villani, 1915. Parte prima, p. H, n. 20]

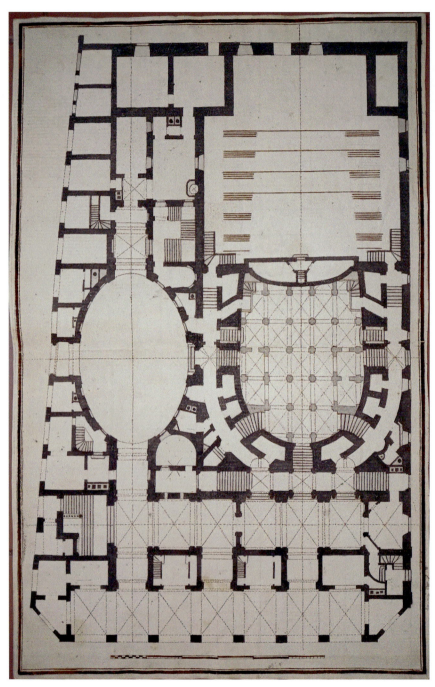

Piante due del Teatro Comunale di Ferrara—sec. XIX [Biblioteca Comunale Ariostea, Fondo Antolini, 70, foglio n.1]

Apparatuses 199

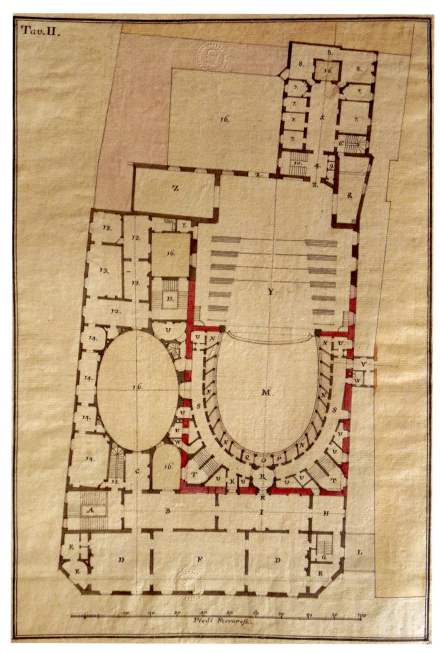

Teatro Comunale di Ferrara [Biblioteca Comunale Ariostea, l Teatro Comunale di Ferrara. Cento anni di storia (1798–1898). Memorie raccolte a cura di Giovanni Pasetti, Tullio Finotti, Luigi Villani, 1915. Parte prima, p. M, n. 26]

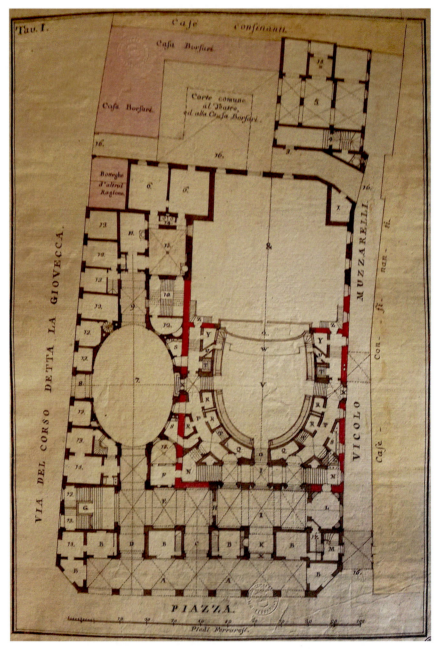

Teatro Comunale di Ferrara [Biblioteca Comunale Ariostea, Il Teatro Comunale di Ferrara. Cento anni di storia (1798–1898). Memorie raccolte a cura di Giovanni Pasetti, Tullio Finotti, Luigi Villani, 1915. Parte prima, p. L, n. 25]

Apparatuses

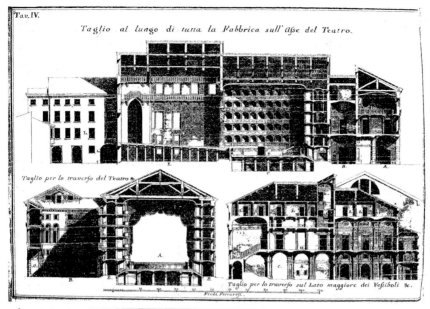

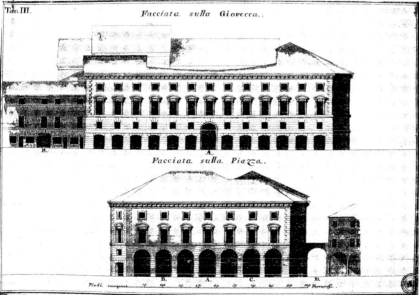

Taglio al lungo di tutta la Fabbrica sull'asse del Teatro, Facciata sulla Giovecca e sulla Piazza [Biblioteca Comunale Ariostea, Il Teatro Comunale di Ferrara. Cento anni di storia (1798–1898). Memorie raccolte a cura di Giovanni Pasetti, Tullio Finotti, Luigi Villani, 1915. Parte prima, p. O, n. 28 e p. N, n. 27]

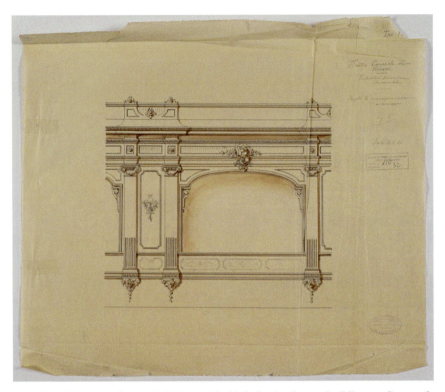

Dettaglio palchi, Teatro Comunale di Ferrara [Archivio Storico Comunale di Ferrara, Cartografia, cart. 1/O]

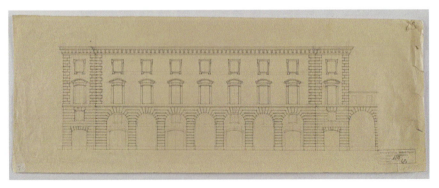

Prospetto del Teatro Comunale [Archivio Storico Comunale di Ferrara, Cartografia, cart. 1/O, n. 45]

Apparatuses

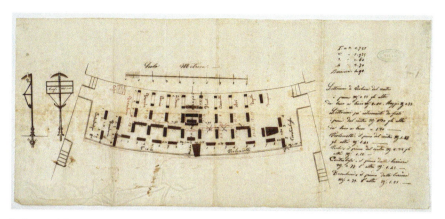

Orchestra del Teatro Comunale [Archivio Storico Comunale di Ferrara, Commissione Pubblici Spettacoli—Direzione Teatrale, Teatro Comunale di Ferrara, Lavori 1852–1885, busta n. 90]

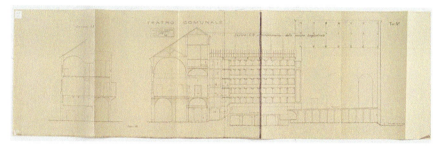

Sezione longitudinale del Teatro Comunale con in evidenza la nuova orchestra ribassata [Archivio Storico Comunale di Ferrara, Cartografia, cart. 1/O, n. 46]

204   Apparatuses

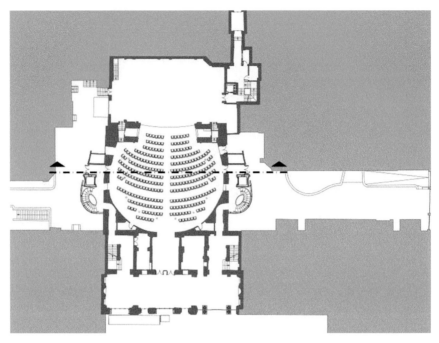

Teatro Carignano, plan [elaboration by the author]

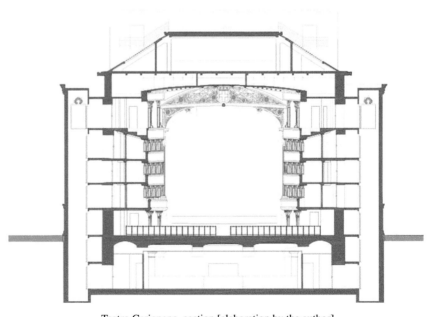

Teatro Carignano, section [elaboration by the author]

Apparatuses

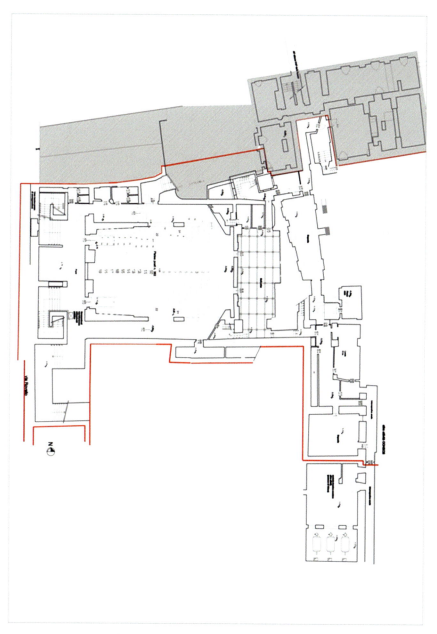

Teatro Piccolo—plan of the basement [Piccolo]

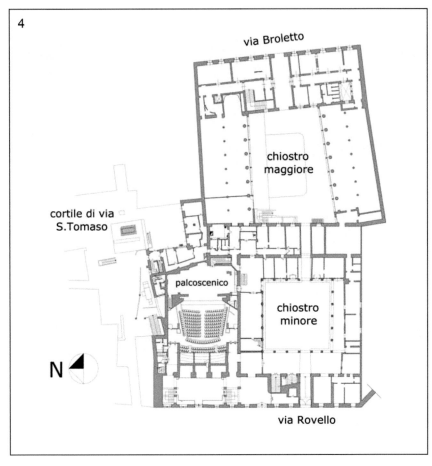

Teatro Piccolo—plan of the groundfloor [Piccolo]

Apparatuses

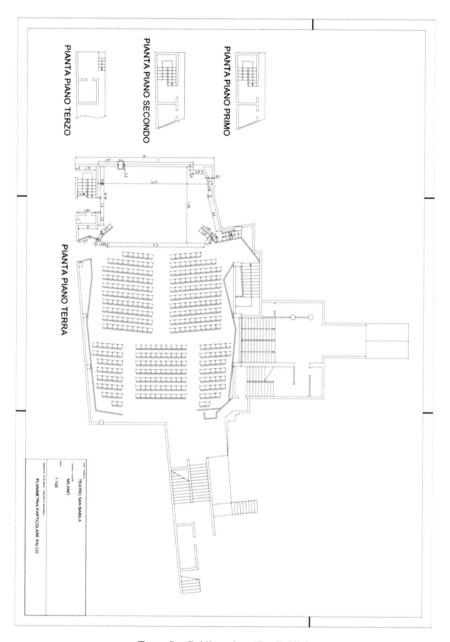

Teatro San Babila—plans [San Babila]

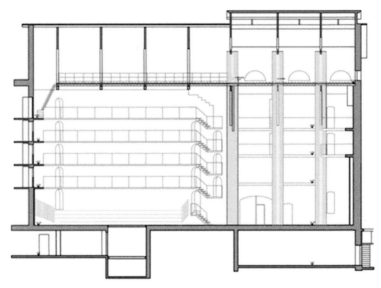

Teatro Studio—section [Piccolo]

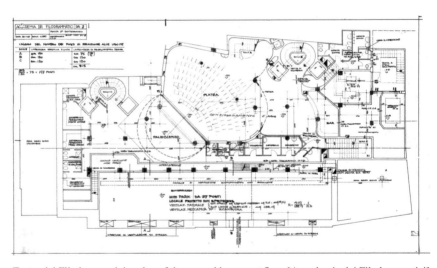

Teatro dei Filodrammatici—plan of the second basement floor [Accademia dei Filodrammatici]

Apparatuses

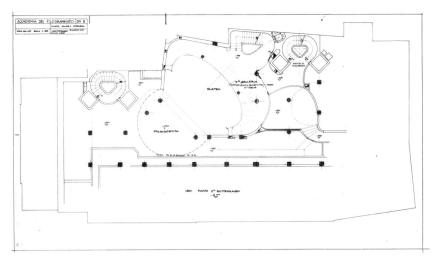

Teatro dei Filodrammatici—plan of the basement [Accademia dei Filodrammatici]

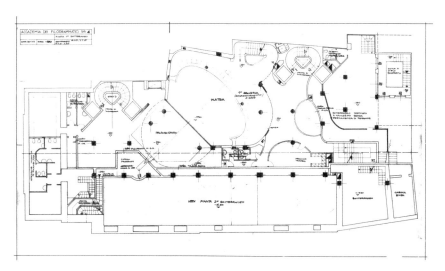

Teatro dei Filodrammatici—plan of the intermediate gallery [Accademia dei Filodrammatici]

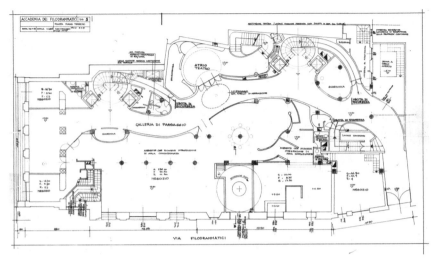

Accademia dei Filodrammatici—plan of the groundflor [Accademia dei Filodrammatici]

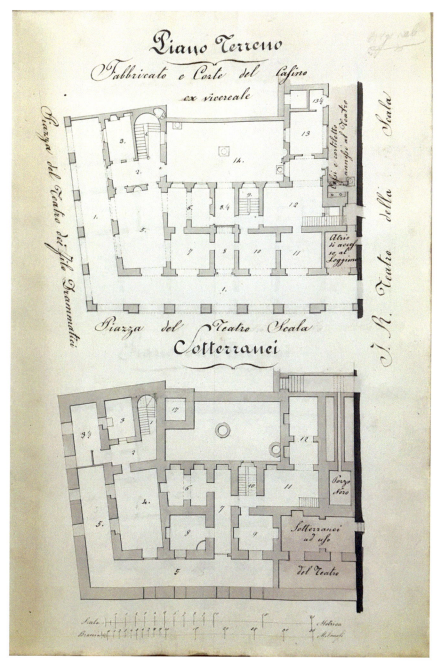

Pianta dei sotterranei e del piano terreno del Casino Ricordi [Archivio di Stato di Milano, Consegna del Casino Erariale annesso all'I.R. Teatro della Scala, Genio Civile di Milano, busta n. 3679]

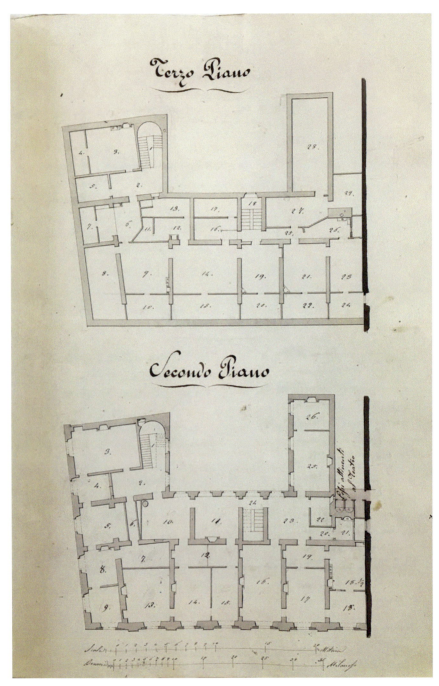

Pianta del secondo e del terzo piano del Casino Ricordi [Archivio di Stato di Milano, Consegna del Casino Erariale annesso all'I.R. Teatro della Scala, Genio Civile di Milano, busta n. 3679]

Apparatuses

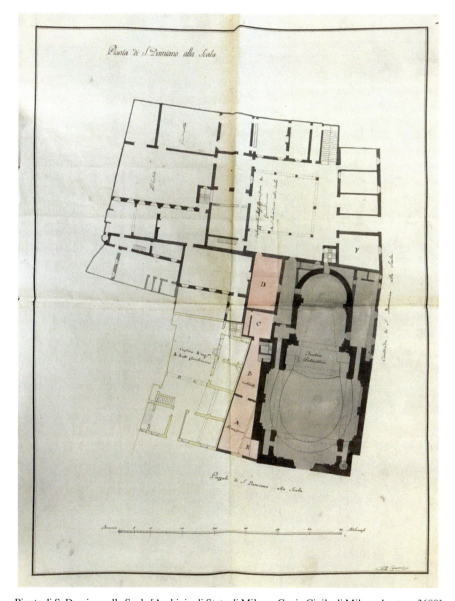

Pianta di S. Damiano alla Scala [Archivio di Stato di Milano, Genio Civile di Milano, busta n. 3680]

Glossary

Organism:	the set of elements linked together by a relationship of necessity that jointly contribute to the same purpose.[1]
Type:	*the heritage of common transmissible characters that precedes the origin of the organism, governing the structure of relations from within, of relations of necessity that inform it.*[2] *The building type corresponds to the persistence of the set of collectively inherited notions, rules and characters, spontaneously or critically accepted and transmitted by a civilized environment in the course of its history.*[3]
Typological process:	*complex of mutations that occurred uniquely in the set of characters of buildings pertaining to different cultural areas and in successive phases.*[4]
Typological migration:	possible outcome of the developing process that involves the transfer from one object to another new and different, for example: from house to temple.
Fence:	primary form of the constructive act, product of the elementary gesture of space appropriation.
Roof:	this is also a primary form of the construction act, the result of the elementary gesture of protection of its own, delimited space.
Nodal axis:	*identifies a centralized geometric place that unifies structure and use of space in a single constructive gesture*[5] to which symbolic value is recognized. The direction of the axis establishes the succession of elements, determines their hierarchy, guides their reading.
Antinodal axis:	peripheral distance.

© Springer Nature Switzerland AG 2022
S. Clemente, *Reconstructing Theatre Architecture*, The Urban Book Series,
https://doi.org/10.1007/978-3-030-89968-4

Polar axis:	vertical axis identified by the center of the building organism, which assumes the unifying function of the pole, at the convergence of the horizontal axes; it determines the position in which the cover will be tied and provides the top–bottom orientation.
Dividing line:	sets the limits of the built environment, describes its boundaries.
Node:	*discontinuity within a continuum, or the intersection of two continuities.*[6]
Serial (antinodal and axial), nodal (uniaxial, biaxial) and polar specialized building:	by specialized building, we mean parts of the construction that do not have residential functions, and which come from housing, nevertheless, at the end of a specialization process.
Overturning:	the paths of the urban fabric are overturned within the specialized organism, determining its origin and developing process.
Knotting:	the constructive gesture of tying together different elements of a structure in order to form a knot.

Notes

1. *«l'insieme di elementi legati tra loro da un rapporto di necessità che concorrono unitariamente a un medesimo fine».* Strappa G (1995) Unità dell'organismo architettonico. edizioni Dedalo, Bari: 24
2. *«il patrimonio di caratteri comuni trasmissibili che precede la formazione dell'organismo governandone dall'interno la struttura di relazioni, di rapporti di necessità che lo informano».* Ibidem: 24
3. *«Il tipo edilizio corrisponde alla persistenza dell'insieme di nozioni, regole e caratteri collettivamente ereditati, spontaneamente o criticamente accettati e trasmessi da un intorno civile nel corso della propria storia».* Ibidem: 32
4. *«complesso delle mutazioni avvenute unitariamente nell'insieme dei caratteri degli edifici pertinenti ad aree culturali diverse e in fasi successive».* Ibidem: 155
5. *«individua un luogo geometrico accentrante che unifica struttura e uso dello spazio in un unico gesto costruttivo».* Ibidem: 82
6. *«discontinuità all'interno di un continuo, o l'intersezione di due continui».* Ibidem: 81

List of Historical Theatres in Italy

© Springer Nature Switzerland AG 2022
S. Clemente, *Reconstructing Theatre Architecture*, The Urban Book Series,
https://doi.org/10.1007/978-3-030-89968-4

Code	Code2	Name	Architect/Engineer	Year	R1	R2	R3	Region	Province	City	Address
.001		Arena del Sole		1810				EmiliaRomagna	Bologna	Bologna	Via dell'Indipendenza, 44
.002		Auditorium Pedrotti	Sebastiano Locati	1885	1907	1972	1995	Marche	Pesaro-Urbino	Pesaro	Piazza Olivieri
.003		Ciema Teatro Olimpia		20th cent				Toscana	Pisa	Vecchiano	Via G.B. Barsuglia, 209
.004		Cine Teatro Minerva		1911	1983	2012		Campania	Napoli	Boscoreale	Via Tenente Angelo Cirillo, 53
.005		Cinema Teatro Aurora		1860	1935			Toscana	Pisa	Montecatini	Via Cesare Battisti
.006		Cinema Teatro Boni		1802				Lazio	Viterbo	Acquapendente	Piazza della Costituente, 9
.007		Cinema Teatro Branzo		1890				Veneto	Padova	Montagnana	Via Matteotti, 100
.008		Cinema Teatro Dante	Leoni, Guidi	1833	1975			Toscana	Arezzo	Sansepolcro	Via XX Settembre
.009		Cinema Teatro Ernesto Rossi		1892				Toscana	Firenze	Signa	Piazza della Repubblica, 1
.010		Cinema Teatro Excelsior		20th cent				Toscana	Firenze	Empoli	Via Cosimo Ridolfi, 75

List of Historical Theatres in Italy

Site of arch. int..	UNESCO Site	ERHT Site	Integrated/Isolated	Construction	Seats	Level	Keyword	Notes	Code
Yes	No	No	Isolated	Mixed structure	888	Regional			.001
Yes	No	No	Integrated	Mixed structure	500	Regional			.002
Yes	No	No	Isolated	Mixed structure	220	Local			.003
Yes	No	No	Integrated	Mixed structure	300	Regional			.004
Yes	No	No	Integrated	Mixed structure	99	Local			.005
Yes	No	No	Isolated	Wooden and brick	250	Local			.006
Yes	No	No	Isolated	Wooden and brick		Local			.007
Yes	No	No	Isolated	Mixed structure	400	Regional			.008
Yes	No	No	Isolated	Wooden and brick		Local			.009
Yes	No	No	Isolated	Mixed structure		Local			.010

Code	Code2	Name	Architect/Engineer	Year	R1	R2	R3	Region	Province	City	Address
.011		Cinema Teatro Iris		1864				Piemonte	Cuneo	Dronero	Via IV Novembre, 9
.012		Cinema Teatro Masaccio		1917	1925	1946	1970	Toscana	Arezzo	San Giovanni Valdarno	Via G. Borsi, 1
.013		Cinema Teatro Masciari		1923				Calabria	Catanzaro	Catanzaro	Piazza Michele Le Pera, 6
.014		Cinema Teatro Palazzo		1932				Puglia	Foggia	Serracapriola	Via Vincenzo Gioberti, 17
.015		Cinema Teatro Russo		20th cent				Calabria	Catanzaro	Lamezia Terme	Via Maggiordomo
.016		Cinema Teatro Tito Schipa		1900				Puglia	Lecce	Gallipoli	Corso Roma, 217
.017		Cinema Teatro Verdi		1920	1932			Puglia	Taranto	Martina Franca	Piazza XX Settembre, 5
.018		Piccolo Teatro		1950				Toscana	Siena	Siena	Via Montanini, 118
.019		Piccolo Teatro degli Instabili		2002	2014			Umbria	Perugia	Assisi	Via Metastasio, 18
.020		Piccolo Teatro della Soffitta		1730				Toscana	Pisa	Pisa	Piazza San Giorgio, 6

List of Historical Theatres in Italy

Site of arch. int..	UNESCO Site	ERHT Site	Integrated/Isolated	Construction	Seats	Level	Keyword	Notes	Code
Yes	No	No	Isolated	Wooden and brick	425	Regional			.011
Yes	No	No	Isolated	Reinforced concrete	468	Local			.012
Yes	No	No	Isolated	Reinforced concrete		Regional			.013
Yes	No	No	Isolated	Reinforced concrete		Local			.014
Yes	No	No	Isolated	Mixed structure		Local		Unusable, to be redeveloped	.015
Yes	No	No	Isolated	Wooden and brick		Local			.016
Yes	No	No	Isolated	Reinforced concrete		Local			.017
Yes	Yes	No	Integrated	Reinforced concrete	100	Local			.018
Yes	No	No	Integrated	Reinforced concrete	100	Local			.019
Yes	No	No	Integrated	Wooden and brick	100	Local			.020

Code	Code2	Name	Architect/ Engineer	Year	R1	R2	R3	Region	Province	City	Address
.021	CODE 5.06	Piccolo Teatro Grassi	Rogers, Zanuso	1947	1952	1999		Lombardia	Milano	Milano	Via Rovello, 2
.022	CODE 5.08	Piccolo Teatro Strehler	Marco Zanuso	1999				Lombardia	Milano	Milano	Largo Greppi, 1
.023	CODE 5.07	Piccolo Teatro Studio	Marco Zanuso	1986				Lombardia	Milano	Milano	Via Rivoli, 6
.024		Politeama Genovese	Dante Datta	1955				Liguria	Genova	Genova	Via Nicolò Bacigalupo, 2
.025		Politeama Pratese	Nervi, Andrè	1925	1954	1999		Toscana	Prato	Prato	Via Giuseppe Garibaldi, 33
.026		Politeama Sociale	Gaetano Malatesta	1912	1933			EmiliaRomagna	Modena	Sassuolo	Via Farosi
.027		Salone Margherita	Giuseppe Pagnani Fusconi	1898	1972	1990		Lazio	Roma	Roma	Via dei Due Macelli, 75
.028		Teatro Gaetano Donizetti	Giovanni Francesco Lucchini	1800	1868	2017		Lombardia	Bergamo	Bergamo	Piazza Cavour, 14
.029		Teatrino	Andrea Busiri Vici	1787	2002			Marche	Ancona	Osimo	Piazza Dante Alighieri, 4
.030		Teatrino Campana		1791				FriuliVeneziaG-iulia	Pordenone	Valvasone	Piazza Castello

List of Historical Theatres in Italy

Site of arch. int.	UNESCO Site	ERHT Site	Integrated/Isolated	Construction	Seats	Level	Keyword	Notes	Code
Yes	No	Yes	Integrated	Mixed structure	488	National	Knotting		.021
Yes	No	Yes	Isolated	Reinforced concrete	968	National	Architectural node		.022
Yes	No	Yes	Integrated	Mixed structure	368	National	Knotting	ex Teatro Fossati	.023
No	No	No	Integrated	Reinforced concrete	1054	National		Theatre of 1868 destroyed in the bombing of 1941, rebuilt	.024
Yes	No	No	Integrated	Reinforced concrete	978	Regional		Opening dome by Nervi	.025
Yes	No	No	Integrated	Reinforced concrete		Local		Inactive since 1964	.026
Yes	Yes	No	Integrated	Mixed structure	450	Regional			.027
Yes	No	No	Isolated	Mixed structure	1154	National		The previous theatre of 1783 was destroyed by a fire in 1797	.028
Yes	No	No	Integrated	Mixed structure	140	Local			.029
Yes	No	No	Integrated	Wooden and brick		Local			.030

Code	Code2	Name	Architect/Engineer	Year	R1	R2	R3	Region	Province	City	Address
.031		Teatrino Civico	Fausto Gozzano	1864				Piemonte	Torino	Chiavasso	Via Defendente Ferrari
.032		Teatrino del Convitto Nazionale Rinaldo Corso		1909				EmiliaRomagna	Reggio Emilia	Correggio	Via Bernieri, 8
.033		Teatrino di Corte	Luigi Canonica	1806				Lombardia	Monza e Brianza	Monza	Viale Brianza, 1
.034		Teatrino di Corte	Ferdinando Fuga	1768	1950	2000		Campania	Napoli	Napoli	Piazza del Plebiscito, 1
.035		Teatrino di Vetriano	Virgilio Biagioni	1890	1997			Toscana	Lucca	Vetraino	Località Carraia, 5
.036		Teatrino di Villa Aldrovandi-Mazzacurati	Francesco Tadolini	1763				EmiliaRomagna	Bologna	Bologna	Via Toscana, 19
.037		Teatrino di Villa Olmo		1797				Lombardia	Como	Como	Via Simone Cantoni, 1
.038		Teatro del Rondò di Bacco	Gaspare Paoletti	1799				Toscana	Firenze	Firenze	Piazza de' Pitti, 1
.039		Teatro 1900		1900				Lombardia	Mantova	Pomponesco	Via Roma, 57
.040		Teatro Accademico		1790	1980			Toscana	Lucca	Bagni di Lucca	Via Umberto 1

List of Historical Theatres in Italy

Site of arch. int.	UNESCO Site	ERHT Site	Integrated/Isolated	Construction	Seats	Level	Keyword	Notes	Code
Yes	No	No	Integrated	Wooden and brick	98	Local		Deconsecrated church and adapted to theatre inside Palazzo del Convento delle Clarisse	.031
Yes	No	No	Integrated	Mixed structure		Local			.032
Yes	No	No	Integrated	Wooden and brick	120	Local			.033
Yes	Yes	No	Integrated	Mixed structure	400	Regional		Partially destroyed during the bombing of 1943	.034
Yes	No	No	Integrated	Mixed structure	95	Local		Philological restoration	.035
Yes	No	Yes	Integrated	Wooden and brick	95	Local		Rebuilt after the bombings of 1943	.036
Yes	No	No	Integrated	Wooden and brick		Local			.037
Yes	Yes	No	Integrated	Wooden and brick		Local			.038
Yes	No	No	Isolated	Brick	144	Local			.039
Yes	No	No	Isolated	Mixed structure	304	Regional			.040

Code	Code2	Name	Architect/ Engineer	Year	R1	R2	R3	Region	Province	City	Address
.041		Teatro Accademico	Francesco Maria Preti	1758	1853	1973	2001	Veneto	Treviso	Castelfranco Veneto	Via Garibaldi
.042		Teatro Adelaide Ristori		1816				FriuliVeneziaGiulia	Udine	Cividale del Friuli	Via Adelaide Ristori, 30
.043		Teatro al Parco	Ugo Pescatori	1939	2000			EmiliaRomagna	Parma	Parma	Viale Piacenza
.044		Teatro Alaleona	Giuseppe Sabbatini	1878				Marche	Fermo	Montegiorgio	Via Roma, 11
.045		Teatro Alessandro Guardassoni	Francesco Gualandi	1879				EmiliaRomagna	Bologna	Bologna	Via Massimo D'Azeglio, 55
.046		Teatro Alfieri	Domenico Svanascini	1860	1979			Piemonte	Asti	Asti	Piazza del Teatro Alfieri, 2
.047		Teatro Alfieri		1860	2006			Toscana	Lucca	Castelnuovo di Carfagna	Via Guglielmo Marconi
.048		Teatro Alfieri	Barnaba Panizza	1855	1949	2002		Piemonte	Torino	Torino	Piazza Solferino, 4
.049		Teatro Alfonso Rendano	Nicola Zumpano	1909	1943			Calabria	Cosenza	Cosenza	Piazza XV Marzo
.050		Teatro Alighieri	T. Meduna, G.B. Meduna	1852	1925	1959	1991	EmiliaRomagna	Ravenna	Ravenna	Via Angelo Mariani, 2

List of Historical Theatres in Italy

Site of arch. int.	UNESCO Site	ERHT Site	Integrated/Isolated	Construction	Seats	Level	Keyword	Notes	Code
Yes	No	No	Isolated	Mixed structure		Regional			.041
Yes	No	No	Isolated	Wooden and brick	438	Regional			.042
Yes	No	No	Isolated	Reinforced concrete	400	Regional			.043
Yes	No	No	Isolated	Brick	300	Regional			.044
Yes	No	No	Integrated	Wooden and brick	124	Local			.045
Yes	No	No	Isolated	Mixed structure	2050	National			.046
Yes	No	No	Isolated	Mixed structure	510	Regional			.047
Yes	No	No	Isolated	Mixed structure	1500	National		Rebuilt following fires in 1860, 1863, 1868, 1927 and the bombings of 1942	.048
Yes	No	No	Isolated	Mixed structure	802	Regional			.049
Yes	No	No	Isolated	Mixed structure	1080	National			.050

228 List of Historical Theatres in Italy

Code	Code2	Name	Architect/Engineer	Year	R1	R2	R3	Region	Province	City	Address
.051		*Teatro alla Scala*	*Piermarini, Botta*	*1778*	*1946*	*2002*		*Lombardia*	*Milano*	*Milano*	*Piazza alla Scala*
.052	*CODE 3.02, 3.12 and 3.24*	*Teatro all'Antica*	*Vincenzo Scamozzi*	*1590*	*1950*	*1964*	*1996*	*Lombardia*	*Mantova*	*Sabbioneta*	*Via del Teatro*
.053		Teatro Ambra Jovinelli	Ulderico Bencivenga	1909	2000			Lazio	Roma	Roma	Via Guglielmo Pepe, 45
.054		Teatro Amiatino		1890	1995			Toscana	Grosseto	Castel del Piano	Piazza Arcipretura, 2
.055		Teatro Amintore Galli	Luigi Poletti	1857	1926	2014		EmiliaRomagna	Rimini	Rimini	Piazza Cavour
.056		Teatro Angelo Dal Foco	Raimondo Compagnini	1752	1981			Marche	Pesaro-Urbino	Pergola	Piazza Garibaldi, 33
.057		Teatro Angelo Mariani		1723	1872	1994		EmiliaRomagna	Ravenna	Sant' Agata Feltria	Piazza Garibaldi, 34

(continued)

(continued)

Code	Code2	Name	Architect/ Engineer	Year	R1	R2	R3	Region	Province	City	Address
.058		Teatro Annibal Caro	Guglielmo Prosperi	1872	1997			Marche	Macerata	Civitanova Alta	Corso Caro Annibal, 2
.059		Teatro Anselmi		1928	1984			Lombardia	Mantova	Pegognaga	Piazza Giacomo Matteotti, 1
.060		Teatro Antonio Belloni		2010				Lombardia	Monza e Brianza	Barlassina	Via Cristoforo Colombo, 38

Site of arch. int.	UNESCO Site	ERHT Site	Integrated/Isolated	Construction	Seats	Level	Keyword	Notes	Code
Yes	*No*	*No*	*Isolated*	*Mixed structure*	*2030*	*International*	*Urban node*	*Destroyed by fire in 1943 and rebuilt*	*.051*
Yes	*Yes*	*Yes*	*Isolated*	*Wooden and brick*		*Local*	*Organism*		*.052*
Yes	Yes	No	Integrated	Reinforced concrete	800	National		Destroyed by fire in 1982 and rebuilt	.053
Yes	No	No	Isolated	Mixed structure	99	Local		Altered in the 1930s and 1940s	.054
Yes	No	No	Isolated	Mixed structure	800	Regional		Damaged by the earthquake of 1916, severely damaged by the bombing of 1943, early Christian basilica was found during the restoration	.055
Yes	No	No	Isolated	Mixed structure	387	Regional			.056
Yes	No	No	Integrated	Mixed structure	99	Local			.057

(continued)

(continued)

Site of arch. int.	UNESCO Site	ERHT Site	Integrated/Isolated	Construction	Seats	Level	Keyword	Notes	Code
Yes	No	No	Isolated	Mixed structure	400	Regional			.058
No	No	No	Isolated	Mixed structure	480	Regional		Unusable after earthquake	.059
No	No	No	Integrated	Reinforced concrete	98	local			.060

Code	Code2	Name	Architect/ Engineer	Year	R1	R2	R3	Region	Province	City	Address
.061		Teatro Apollo		1900				Sicilia	Trapani	Castellammare del Golfo	Corso Bernardo Mattarella, 15
.062		Teatro Apollo	Tassoni	1926	2008			Puglia	Lecce	Lecce	Via Salvatore Trinchese, 13A
.063		Teatro Apollo	Francesco Luciani	1839	1907	1984		Marche	Macerata	Mogliano	Piazza Garibaldi
.064		Teatro Apollo		1887				Marche	Ancona	Mondavio	Piazza Giovanni della Rovere
.065		Teatro Argentina	Theodoli, Portoghesi	1732	1993			Lazio	Roma	Roma	Largo di Torre Argentina, 52
.066		Teatro Ariosto	Achille Grimaldi	1878	1927	1993		EmiliaRomagna	Reggio Emilia	Reggio Emilia	Corso Cairoli, 1
.067	CODE 4.21	Teatro Arnoldi	Arnoldi	1883				Lombardia	Mantova	Mantova	Via Mainolda

(continued)

(continued)

Code	Code2	Name	Architect/Engineer	Year	R1	R2	R3	Region	Province	City	Address
.068		Teatro Asioli	Francesco Cipriano Forti	1873	1890	1973	2002	EmiliaRomagna	Reggio Emilia	Correggio	Corso C. B. Conte di Cavour, 9
.069		Teatro Astra	Giacomo Bauto	1811				Veneto	Vicenza	Bassano del Grappa	Viale Martiri, 46
.070		Teatro Augusto Massari		1856	1980			EmiliaRomagna	Ravenna	San Giovanni in Marignano	Via Serpieri

Site of arch. int.	UNESCO Site	ERHT Site	Integrated/Isolated	Construction	Seats	Level	Keyword	Notes	Code
No	No	No	Isolated	Reinforced concrete	198	Local			.061
Yes	No	No	Isolated	Reinforced concrete	705	Regional			.062
Yes	No	No	Isolated	Mixed structure	210	Local			.063
Yes	No	No	Integrated	Wooden and brick	99	Local			.064
Yes	Yes	No	Isolated	Mixed structure	720	National			.065
Yes	No	No	Isolated	Mixed structure	780	Regional		The previous theatre of 1741 was destroyed by a fire in 1851	.066
Yes	*Yes*	*No*	*Integrated*	*Mixed structure*		*Regional*	*Recast*		*.067*
Yes	No	No	Isolated	Mixed structure	499	Regional		Damaged by fires in 1889 and 1909, and by earthquake in 1996	.068
Yes	No	No	Isolated	Wooden and brick		Local			.069
Yes	No	No	Isolated	Mixed structure	110	Local			.070

List of Historical Theatres in Italy

Code	Code2	Name	Architect/Engineer	Year	R1	R2	R3	Region	Province	City	Address
.071		Teatro Aurora		1930	1995			Toscana	Firenze	Montelupo Fiorentino	Piazza San Rocco
.072		Teatro Aventino		1848	1938	1998		Abruzzo	Chieti	Palena	Corso Umberto I
.073		Teatro Aycardi		1803				Liguria	Savona	Finale Ligure	Piazza Aicardi
.074		Teatro Battelli	Dante Giampaoli	1932	1984			Marche	Pesaro-Urbino	Macerata Feltria	Via Pitino
.075		Teatro Belli		1880	1972	1984	2010	Lazio	Roma	Roma	Piazza di Sant'Apollonia, 11
.076		Teatro Bellini	Carmelo Sciuto Patti	1870				Sicilia	Catania	Acireale	Via Romeo, 63
.077		Teatro Bellini	Sada, Scala	1890				Sicilia	Catania	Catania	Via Giuseppe Perrotta, 12
.078		Teatro Bellini	Carlo Sorgente	1878				Campania	Napoli	Napoli	Via Conte di Ruvo, 14
.079		Teatro Bellini		1742	1808	1840	1965	Sicilia	Palermo	Palermo	Piazza Bellini, 1
.080		Teatro Beniamino Gigli		1816	1920	1930	1979	Marche	Ancona	Monte Roberto	Via G. Marconi

Site of arch. int.	UNESCO Site	ERHT Site	Integrated/Isolated	Construction	Seats	Level	Keyword	Notes	Code
No	No	No	Isolated	Reinforced concrete	200	Local			.071
Yes	No	No	Isolated	Mixed structure	99	Local			.072
Yes	No	No	Isolated	Wooden and brick	250	Local			.073
Yes	No	No	Isolated	Mixed structure	400	Regional			.074
Yes	Yes	No	Isolated	Mixed structure	140	Local		Previously inside a convent	.075
Yes	No	No	Isolated	Mixed structure	600	Regional		Damaged by fire in 1952	.076
Yes	No	No	Isolated	Wooden and brick	1200	National			.077
Yes	Yes	No	Isolated	Wooden and brick	940	National			.078
Yes	No	No	Isolated	Mixed structure	274	Local		Damaged by fire in 1965 and restored	.079
Yes	No	No	Isolated	Mixed structure	115	Local		Damaged by the snowfall of 1962	.080

Code	Code2	Name	Architect/Engineer	Year	R1	R2	R3	Region	Province	City	Address
.081		Teatro Bertagnolio	Bonaventura Bertagnolio	1897				Piemonte	Torino	Chiaverano	Via del Teatro, 19
.082		Teatro Besostri		1892	1954	2006		Lombardia	Pavia	Pavia Mede	Via Giuseppe Besostri, 33
.083		Teatro Biondo	Nicolò Mineo	1903				Sicilia	Palermo	Palermo	Via Roma, 258
.084		Teatro Boccaccio		20th cent				Toscana	Firenze	Certaldo	Via del Castello, 2
.085		*Teatro Bonacossi*	*Giovanni Tosi*	*1662*	*1840*	*1922*		*EmiliaRomagna*	*Ferrara*	*Ferrara*	*Via del Turco*
.086		Teatro Bonci	Vincenzo Ghinelli	1843				EmiliaRomagna	Forli-Cesena	Cesena	Piazza Guidazzi, 9
.087		Teatro Borgatti	Antonio Giordani	1856	1972	2019		EmiliaRomagna		Cento	Corso Guercino
.088		Teatro Bramante	Ercole Salmi	1869	1949			Marche	Pesaro-Urbino	Urbania	Piazza San Cristoforo
.089		Teatro Brancaccio	Carlo Sacconi	1916				Lazio	Roma	Roma	via Merulana 244
.090		Teatro Brendel		1865				Toscana	Firenze	Firenze	Via delle Casine, 21

Site of arch. int.	UNESCO Site	ERHT Site	Integrated/Isolated	Construction	Seats	Level	Keyword	Notes	Code
Yes	No	No	Isolated	Wooden and brick		Local			.081
Yes	No	No	Isolated	Mixed structure	340	Regional			.082
Yes	No	No	Integrated	Reinforced concrete		Regional			.083
Yes	No	No	Isolated	Wooden and brick		Regional			.084
Yes	*Yes*	*No*	*Integrated*	*Reinforced concrete*		*Regional*	*Organism*	*Change of intended use*	*.085*
Yes	No	No	Isolated	Brick	800	Regional			.086
Yes	No	No	Isolated	Mixed structure	434	Regional			.087
Yes	No	No	Isolated	Mixed structure	324	Regional			.088
Yes	Yes	No	Isolated	Reinforced concrete	1300	National			.089
Yes	Yes	No	Isolated	Wooden and brick		Local			.090

List of Historical Theatres in Italy

Code	Code2	Name	Architect/Engineer	Year	R1	R2	R3	Region	Province	City	Address
.091		Teatro Cagnoni		1873				Lombardia	Pavia	Vigevano	Corso Vittorio Emanuele II, 43
.092		Teatro Caio Melisso	Ireneo Aleandri	1850	1959			Umbria	Perugia	Spoleto	Piazza del Duomo, 1
.093		Teatro Cantero	Ido Gazzano	1926				Liguria	Genova	Chiavari	Piazza Giacomo Matteotti, 23
.094	CODE 2.05	Teatro Capitolino	Pietro Rosselli	1513				Lazio	Roma	Roma	
.095		Teatro Caporali	Francesco Tarducci	1786	1856	1921		Umbria	Perugia	Panicale	Via Cesare Caporali, 16
.096		Teatro Capranica	Carlo Buratti	1679				Lazio	Roma	Roma	Piazza Capranica, 101
.097		Teatro Carbonetti		1881	1963			Lombardia	Pavia	Broni	Via Leonardo Da Vinci, 27
.098		Teatro Carcano	Luigi Canonica	1803	2010			Lombardia	Milano	Milano	Corso di Porta Romana, 63
.099	CODE 6.01 and 6.21	Teatro Carlo Felice	Barabino, Scarpa, Rossi	1828	1991			Liguria	Genova	Genova	Passo Eugenio Montale, 4
.100		Teatro Carlo Goldoni	Buffoni, Fellini, Pasqui	1863	1963	2005		Marche	Ancona	Corinaldo	Via del Corso, 66

Site of arch. int.	UNESCO Site	ERHT Site	Integrated/Isolated	Construction	Seats	Level	Keyword	Notes	Code
Yes	No	No	Isolated	Wooden and brick	602	Regional			.091
Yes	No	No	Isolated	Mixed structure	300	Regional	Knotting		.092
Yes	No	No	Integrated	Reinforced concrete		Regional			.093
Yes	Yes	No	Isolated	Wooden	1800	National	Enclosure	Temporary	.094
Yes	No	No	Isolated	Mixed structure		Regional			.095
Yes	Yes	No	Isolated	Wooden and brick	800	Regional		Closed in 1881, it has been a congress center since 2005	.096
Yes	No	No	Isolated	Mixed structure	362	Regional			.097
Yes	No	No	Integrated	Mixed structure	990	Regional			.098
Yes	No	No	Integrated	Mixed structure	2000	International	Urban node		.099
Yes	No	No	Integrated	Mixed structure		Regional			.100

List of Historical Theatres in Italy

Code	Code2	Name	Architect/Engineer	Year	R1	R2	R3	Region	Province	City	Address
.101		Teatro Carlo Goldoni	Giuseppe Cappellini	1847	1997			Toscana	Livorno	Livorno	Via Goldoni, 83
.102		Teatro Carlo Marenco		1861	1971			Piemonte	Cuneo	Ceva	Via C. Francesco Adriano, 1
.103		Teatro Castagnoli	Augusto Corbi	1892	1982			Toscana	Grosseto	Scansano	Via XX Settembre, 15
.104		Teatro Cavour	Nicolò Arnaldi	1862	1953	1983		Liguria	Imperia	Imperia	Via Felice Cascione, 35
.105		Teatro Cecilia Gallerani		1923	1983			Lombardia	Cremona	San Giovanni in Croce	Via Grasselli Barni
.106		Teatro Cicconi	Ireneo Aleandri	1872	1954			Marche	Fermo	Sant'Elpido a Mare	Corso Baccio, 82
.107		Teatro Cielo d'Alcamo		1961				Sicilia	Trapani	Alcamo	Piazza Castello
.108		Teatro Cinema Apollo	Sesto Baccarini	1914				EmiliaRomagna	Forlì-Cesena	Forlì	Via Mentana, 8
.109		Teatro Cinema Giuseppe Sarti	Vincenzo Pritelli	1910				EmiliaRomagna	Ravenna	Faenza	Via Carlo Cesare Scaletta, 10
.110		Teatro Cittadino		1869	2006			Puglia	Bari	Noicattaro	Via Carmine, 76

Site of arch. int.	UNESCO Site	ERHT Site	Integrated/Isolated	Construction	Seats	Level	Keyword	Notes	Code
Yes	No	No	Isolated	Mixed structure	900	Regional			.101
Yes	No	No	Isolated	Mixed structure	300	Regional			.102
Yes	No	No	Isolated	Mixed structure	200	Local			.103
Yes	No	No	Integrated	Mixed structure	634	Regional		Unusable, to be redeveloped	.104
Yes	No	No	Isolated	Reinforced concrete	223	Local			.105
Yes	No	No	Isolated	Mixed structure		Local			.106
Yes	No	No	Isolated	Reinforced concrete	808	Regional		Built after the demolition of the Ferrigno Theatre	.107
Yes	No	No	Isolated	Reinforced concrete	214	Local		Currently closed	.108
Yes	No	No	Isolated	Reinforced concrete	350	Regional			.109
Yes	No	No	Isolated	Mixed structure	200	Local			.110

Code	Code2	Name	Architect/Engineer	Year	R1	R2	R3	Region	Province	City	Address
.111		Teatro Civico	Franco Poggi	1862				Sardegna	Sassari	Alghero	Piazza del Teatro, 7
.112		Teatro Civico	Giuseppe Cominotti	1836	2006			Sardegna	Cagliari	Cagliari	Via Mario de Candia
.113		Teatro Civico		1830	1932	1994		Campania	Caserta	Caserta	Via Giuseppe Mazzini, 71
.114		Teatro Civico	Antonio Giordani	1871	2004			EmiliaRomagna	Bologna	Crevalcore	Via G. Matteotti, 106
.115		Teatro Civico	Ippolito Cremona	1846	1933	1989		Liguria	La Spezia	La Spezia	Piazza Mentana, 1
.116		Teatro Civico	Domenico Mollaioli	1872	1995			Umbria	Perugia	Norcia	Piazza Vittorio Veneto
.117		Teatro Civico	Giuseppe Cominotti	1829	1967			Sardegna	Sassari	Sassari	Corso Vittorio Emanuele II
.118		Teatro Civico	Ferruccio Chemello	1909	1993			Veneto	Vicenza	Schio	Via Pozza Maraschin, 19
.119		Teatro Civico		1868	1933			Piemonte	Torino	Susa	Vicolo Gran Valentino
.120		Teatro Civico		1836				Piemonte	Alessandria	Tortona	Via Ammiraglio Mirabello, 3

Site of arch. int.	UNESCO Site	ERHT Site	Integrated/Isolated	Construction	Seats	Level	Keyword	Notes	Code
Yes	No	No	Isolated	Wooden and brick	284	Local			.111
Yes	No	No	Isolated	Mixed structure	100	Local			.112
Yes	No	No	Isolated	Mixed structure		Regional			.113
Yes	No	No	Isolated	Mixed structure	320	Regional		Reconstruction of the previous theatre of 1729 demolished in 1886, damaged by the 2012 earthquake	.114
Yes	No	No	Isolated	Mixed structure	940	Regional		Current building altered in 1933 by Oliva	.115
Yes	No	No	Isolated	Mixed structure		Local		Interior destroyed by fire in 1952, unusable after the 2016 earthquake	.116
Yes	No	No	Isolated	Mixed structure	290	Local			.117
Yes	No	No	Isolated	Mixed structure	335	Regional			.118
Yes	No	No	Isolated	Mixed structure		Local		Damaged by a fire of 1899	.119
Yes	No	No	Isolated	Wooden and brick	475	Regional			.120

List of Historical Theatres in Italy

Code	Code2	Name	Architect/Engineer	Year	R1	R2	R3	Region	Province	City	Address
.121		Teatro Civico	Daniele Donghi	1901				Piemonte	Vercelli	Varallo	Piazza Vittorio Emanuele II
.122		Teatro Civico	Capponi, Sappia	1903	2007			Liguria	Imperia	Ventimiglia	Via Aprosio
.123		Teatro Civico	Bezzi, Marini	1878	1983			Piemonte	Asti	Moncalvo	Piazza Garibaldi, 28
.124		Teatro Civico Comunale		19th cent				Piemonte	Cuneo	Busca	Piazza del Teatro
.125		Teatro Civico Ebe Stignani	Giuseppe Magistretti	1812	1855	1974	2010	EmiliaRomagna	Bologna	Imola	Via Giuseppe Verdi, 1/3
.126		Teatro Civico Milanollo	Maurizio Eula	1836				Piemonte	Cuneo	Savigliano	Piazza Casimiro Turletti, 7
.127		Teatro Civico Toselli	Carlo Barabino	1803	1920	1968	1996	Piemonte	Cuneo	Cuneo	Via Teatro Giovanni Toselli, 9
.128		Teatro Civico Vittorio Emanuele II	Celestino Braccio	1846				Lombardia	Pavia	Mortara	Via Vittorio Emanuele II
.129		Teatro Clitunno	Domenico Mollaioli	1877	1932	1955	1987	Umbria	Perugia	Trevi	Via del Teatro
.130		Teatro Coccia	Giuseppe Oliviero	1886				Piemonte	Novara	Novara	Via Fratelli Rosselli, 47

Site of arch. int.	UNESCO Site	ERHT Site	Integrated/Isolated	Construction	Seats	Level	Keyword	Notes	Code
Yes	No	No	Isolated	Mixed structure		Regional			.121
Yes	No	No	Isolated	Mixed structure	402	Regional			.122
Yes	No	No	Isolated	Mixed structure	350	Regional			.123
No	No	No	Isolated	Mixed structure	149	Local			.124
Yes	No	No	Isolated	Mixed structure	468	Regional			.125
Yes	No	No	Isolated	Wooden and brick	352	Regional			.126
Yes	No	No	Isolated	Mixed structure	580	Regional			.127
Yes	No	No	Isolated	Wooden and brick	500	Local			.128
Yes	No	No	Isolated	Mixed structure	220	Local			.129
Yes	No	No	Isolated	Wooden and brick	917	Regional			.130

List of Historical Theatres in Italy

Code	Code2	Name	Architect/Engineer	Year	R1	R2	R3	Region	Province	City	Address
.131		Teatro Colombo		1913	1997			Toscana	Lucca	Valdottavo	Piazza Tricolore, 30
.132		Teatro Comunale	Raffaele Fiordalisi	1872				Puglia	Bari	Acquaviva delle Fonti	piazza Vittorio Emanuele
.133		Teatro Comunale	Gianbattista Scarpari	1935				Veneto	Rovigo	Adria	Piazza Cavour
.134		Teatro Comunale		19th cent				Lazio	Frosinone	Alvito	Piazza Guglielmo Marconi, 1
.135		Teatro Comunale		20th cent	1994			Veneto	Verona	Asparetto	Piazza Alessandro Manzoni, 1
.136		Teatro Comunale		1911	1984			Abruzzo	Chieti	Atessa	Largo Municipio, 1
.137		Teatro Comunale	Nicola Mezucelli	1881	1946	1962	1980	Abruzzo	Teramo	Atri	Piazza Duchi d'Acquaviva
.138		Teatro Comunale		1814	1855	1970	2007	Veneto	Rovigo	Badia Polesine	Piazza Vittorio Emanuele
.139		Teatro Comunale		1890	1921	1933	1999	Toscana	Firenze	Bagno a Ripoli	Via Montisoni, 10
.140		Teatro Comunale	Giuseppe Segusini	1835	1866	1948	1993	Veneto	Belluno	Belluno	Piazza dei Martiri

Site of arch. int.	UNESCO Site	ERHT Site	Integrated/Isolated	Construction	Seats	Level	Keyword	Notes	Code
Yes	No	No	Isolated	Mixed structure	170	Local			.131
Yes	No	No	Isolated	Wooden and brick		Local			.132
Yes	No	No	Isolated	Reinforced concrete	764	Regional			.133
No	No	No	Integrated	Mixed structure	120	Local			.134
No	No	No	Integrated	Mixed structure	100	Local			.135
Yes	No	No	Isolated	Reinforced concrete	250	Local			.136
Yes	No	No	Isolated	Mixed structure	300	Regional			.137
Yes	No	No	Isolated	Mixed structure		Local			.138
Yes	No	No	Isolated	Mixed structure	99	Local			.139
Yes	No	No	Isolated	Mixed structure	352	Regional			.140

List of Historical Theatres in Italy   249

Code	Code2	Name	Architect/Engineer	Year	R1	R2	R3	Region	Province	City	Address
.141		Teatro Comunale		1900	1980	1993		Sicilia	Catania	Belpasso	Via XII Traversa, 95
.142		Teatro Comunale	Antonio Galli Bibiena	1754	1820	1854	1936	EmiliaRomagna	Bologna	Bologna	Largo Respighi, 1
.143		Teatro Comunale	Giuseppe Mascolini	1832				EmiliaRomagna	Ravenna	Brisighella	Via Naldi, 2
.144		Teatro Comunale		1860	2001			Sicilia	Catania	Bronte	Piazza Spedalieri, 4
.145		Teatro Comunale		1892				Lazio	Viterbo	Canino	Via Roma, 2
.146		Teatro Comunale		1890	1988	2016		Toscana	Pisa	Capannoli	Piazza del Popolo
.147		Teatro Comunale	Cesare Costa	1861	1939	1978	2013	EmiliaRomagna	Modena	Carpi	Piazza Martiri, 72
.148		Teatro Comunale		1906	1987			Toscana	Pisa	Casale Marittimo	Via Roma, 50
.149		Teatro Comunale	Bertelli, Perticucci	1911	2000			Toscana	Arezzo	Castiglion Fiorentino	Via Trieste, 7
.150		Teatro Comunale		20th cent				TrentinoAltoAdige	Trento	Cavalese	Via Roma, 9

Site of arch. int.	UNESCO Site	ERHT Site	Integrated/Isolated	Construction	Seats	Level	Keyword	Notes	Code
Yes	No	No	Isolated	Mixed structure	500	Regional			.141
Yes	No	Yes	Isolated	Mixed structure	1034	National			.142
Yes	No	No	Isolated	Wooden and brick	200	Local		Damaged during the war, restored in the 1950s	.143
Yes	No	No	Isolated	Mixed structure	230	Local			.144
Yes	No	No	Isolated	Wooden and brick	400	Regional			.145
Yes	No	No	Integrated	Mixed structure	110	Local			.146
Yes	No	No	Isolated	Mixed structure	582	Regional			.147
No	No	No	Isolated	Mixed structure	80	Local			.148
Yes	No	No	Isolated	Reinforced concrete	335	Regional			.149
No	No	No	Isolated	Mixed structure		Local			.150

List of Historical Theatres in Italy

Code	Code2	Name	Architect/Engineer	Year	R1	R2	R3	Region	Province	City	Address
.151		Teatro Comunale	Augusto Nencioni	1919	1926	1994		Toscana	Arezzo	Cavriglia	Piazza Enrico Berlinguer, 1
.152		Teatro Comunale	Antonio Caruso	1818				Sicilia	Palermo	Cefalù	Via Salvatore Spinuzza, 115
.153		Teatro Comunale	Antonio Guariglia	1878	1983			Puglia	Brindisi	Ceglie Messapica	Via S. Rocco, 1
.154		Teatro Comunale	Luigi Poletti	1860	1875	1894	1991	EmiliaRomagna	Reggio Emilia	Cervia	Via XX Settembre, 125
.155		Teatro Comunale		1865	1981			EmiliaRomagna	Forli-Cesena	Cesenatico	Via Mazzini, 10–12
.156		Teatro Comunale		1854	1997			Marche	Ancona	Chiaravalle	Corso Giacomo Matteotti, 116
.157		Teatro Comunale	Emilio Giampietro	1856	2002			Abruzzo	Pescara	Città Sant'Angelo	Piazza IV Novembre
.158		Teatro Comunale		1810				Lazio	rieti	Collevecchio	Piazza Statuto, 5
.159		Teatro Comunale		1875				Veneto	Verona	Cologna Veneta	Via Papesso, 10
.160		Teatro Comunale		1894	1934			EmiliaRomagna	Bologna	Conselice	Via Selice, 125

Site of arch. int.	UNESCO Site	ERHT Site	Integrated/Isolated	Construction	Seats	Level	Keyword	Notes	Code
Yes	No	No	Isolated	Mixed structure	490	Regional		Damaged by fire in 1921	.151
Yes	No	No	Isolated	Wooden and brick		Local			.152
Yes	No	No	Isolated	Mixed structure	382	Regional			.153
Yes	No	No	Isolated	Mixed structure	230	Local			.154
Yes	No	No	Isolated	Mixed structure	271	Local			.155
Yes	No	No	Isolated	Mixed structure	300	Local			.156
Yes	No	No	Isolated	Mixed structure	200	Local			.157
Yes	No	No	Isolated	Wooden and brick	120	Local			.158
Yes	No	No	Isolated	Wooden and brick		Local			.159
Yes	No	No	Isolated	Mixed structure	400	Regional			.160

List of Historical Theatres in Italy

Code	Code2	Name	Architect/Engineer	Year	R1	R2	R3	Region	Province	City	Address
.161		Teatro Comunale		1908				FriuliVeneziaGiulia	Gorizia	Cormons	Via Nazario Sauro, 17
.162		Teatro Comunale	Pier Giuseppe Zerboglio	1866				Piemonte	Torino	Cuorgnè	Via Garibaldi, 17
.163		Teatro Comunale		1790	1960			Marche	Macerata	Esanatoglia	Piazza Giacomo Leopardi, 1
.164		Teatro Comunale	Telemaco Bonaiuti	1862	1958			Toscana	Firenze	Firenze	Corso Italia, 16
.165		Teatro Comunale		1910				Lazio	Frosinone	Fiuggi	Piazza Trento e Trieste
.166		Teatro Comunale	Soncini, Montacchini	1866	1949	1965	2008	EmiliaRomagna	Parma	Fontanellato	Via Sanvitale, 21
.167		Teatro Comunale		18th cent				Sicilia	Messina	Furnari	Via Palermo
.168		Teatro Comunale		1929	1981			EmiliaRomagna	Forlì-Cesena	Galeata	Via Cenni, 10
.169		Teatro Comunale		1913				EmiliaRomagna	Forlì-Cesena	Gambettola	Piazza II Risorgimento, 6
.170		Teatro Comunale		1890	1950			Lombardia	Mantova	Goito	Piazza Antonio Gramsci, 5

Site of arch. int.	UNESCO Site	ERHT Site	Integrated/Isolated	Construction	Seats	Level	Keyword	Notes	Code
No	No	No	Isolated	Mixed structure		Local			.161
Yes	No	No	Integrated	Mixed structure		Local			.162
Yes	No	No	Isolated	Mixed structure	180	Local			.163
Yes	No	No	Isolated	Mixed structure	1800	National			.164
Yes	No	No	Integrated	Reinforced concrete	300	Regional			.165
Yes	No	No	Isolated	Mixed structure	200	Local			.166
No	No	No	Isolated	Wooden and brick	500	Regional			.167
No	No	No	Isolated	Reinforced concrete	177	Local			.168
Yes	No	No	Isolated	Reinforced concrete	100	Local			.169
Yes	No	No	Isolated	Mixed structure		Local			.170

List of Historical Theatres in Italy

Code	Code2	Name	Architect/ Engineer	Year	R1	R2	R3	Region	Province	City	Address
.171		Teatro Comunale		1900	1932	1999		Lombardia	Mantova	Gonzaga	Via Martiri di Belfiore
.172		Teatro Comunale		1870				Marche	Pesaro-Urbino	Gradara	Via Zanvettori
.173		Teatro Comunale	Gian Battista Fattori	1771	1905	1975		EmiliaRomagna	Reggio Emilia	Gualtieri	Piazza IV Novembre
.174		Teatro Comunale		1890	1958	2001		Toscana	Pisa	Lajatico	Via Garibaldi, 5
.175		Teatro Comunale		1929	1937	2007		Toscana	Pistoia	Lamporecchio	Via della Costituzione, 11
.176		Teatro Comunale	Giovanni Carraro	1892	1982			Veneto	Vicenza	Lonigo	Piazza Giacomo Matteotti, 1
.177		Teatro Comunale	Michele Campanella	1850				Puglia	Taranto	Massafra	Piazza Garibaldi, 10
.178		Teatro Comunale		1850				Lombardia	Mantova	Medole	Piazza Castello
.179		Teatro Comunale		1894	1930	2000		Puglia	Brindisi	Mesagne	Via Federico II
.180		Teatro Comunale	Raffaele Grilli	1877	1980			Marche	Ancona	Montecarotto	Piazza del Teatro

Site of arch. int.	UNESCO Site	ERHT Site	Integrated/Isolated	Construction	Seats	Level	Keyword	Notes	Code
Yes	No	No	Isolated	Mixed structure		Local			.171
Yes	No	No	Isolated	Wooden and brick		Local			.172
Yes	No	No	Integrated	Mixed structure		Local			.173
Yes	No	No	Isolated	Mixed structure	99	Local			.174
No	No	No	Isolated	Reinforced concrete		Local			.175
Yes	No	No	Isolated	Mixed structure	1000	National			.176
Yes	No	No	Isolated	Wooden and brick		Local			.177
Yes	No	No	Integrated	Wooden and brick	70	Local			.178
Yes	No	No	Isolated	Mixed structure	213	Local			.179
Yes	No	No	Isolato	Mixed structure	148	Local			.180

List of Historical Theatres in Italy

Code	Code2	Name	Architect/Engineer	Year	R1	R2	R3	Region	Province	City	Address
.181		Teatro Comunale		1890	1980			EmiliaRomagna	Bologna	Mordano	Via S.Eustachio, 22
.182		Teatro Comunale		1871	2005			Marche	Macerata	Morrovalle	Piazza Vittorio Emanuele, 2
.183		Teatro Comunale	Quintino Tarantino	1908				Puglia	Lecce	Nardò	Corso Vittorio Emanuele II, 20
.184		Teatro Comunale	Oronzo Bernardini	1891	1950	1980		Puglia	Lecce	Novoli	Piazza Regina Margherita, 12
.185		Teatro Comunale	Andrea Bibolotti	1784	1937	1960	1991	Toscana	Lucca	Pietrasanta	Piazza Duomo, 13
.186		Teatro Comunale		1856	1954			EmiliaRomagna	Bologna	Pieve di Cento	Piazza Andrea Costa, 17
.187		Teatro Comunale		1881				Puglia	Bari	Putignano	Corso Umberto I
.188		Teatro Comunale		20th cent				Lombardia	Mantova	Quingentole	Piazza Italia
.189		Teatro Comunale	Luigi Sottili	1838				EmiliaRomagna	Reggio Emilia	Reggiolo	Piazza Martiri
.190		Teatro Comunale		1911	1993			EmiliaRomagna	Reggio Emilia	Rio Saliceto	Via Giuseppe Garibaldi, 28

Site of arch. int.	UNESCO Site	ERHT Site	Integrated/Isolated	Construction	Seats	Level	Keyword	Notes	Code
Yes	No	No	Integrated	Mixed structure	99	Local			.181
Yes	No	No	Isolated	Mixed structure	170	Local			.182
No	No	No	Isolated	Mixed structure	250	Local			.183
Yes	No	No	Isolated	Mixed structure	200	Local			.184
Yes	No	No	Isolated	Mixed structure	540	Regional			.185
Yes	No	No	Isolated	Mixed structure	145	Local			.186
Yes	No	No	Isolated	Mixed structure		Local			.187
No	No	No	Isolated	Mixed structure	170	Local		Damaged by the 2012 earthquake	.188
Yes	No	No	Isolated	Mixed structure		Local		Unusable due to the 2012 earthquake	.189
Yes	No	No	Isolated	Reinforced concrete	210	Local			.190

List of Historical Theatres in Italy

Code	Code2	Name	Architect/Engineer	Year	R1	R2	R3	Region	Province	City	Address
.191		Teatro Comunale		20th cent				EmiliaRomagna	Ravenna	Riolo Terme	Corso Matteotti 26
.192		Teatro Comunale		1790	1988	2002		Lombardia	Pavia	Rivanazzano Terme	Viale Europa
.193		Teatro Comunale		1887	1924	1995		EmiliaRomagna	Ravenna	Russi di Romagna	Via Cavour
.194		Teatro Comunale	Arturo Prati	1907	1985			EmiliaRomagna	Modena	San Felice sul Panaro	Via Mazzini, 12
.195		Teatro Comunale	Pier Luigi Montacchini	1857				EmiliaRomagna	Parma	San Secondo Parmense	Via Giuseppe Garibaldi, 24
.196		Teatro Comunale	Giuseppe Damiani Almeyda	1897	1952	2002	2010	Sicilia	Siracusa	Siracusa	Via del Teatro
.197		Teatro Comunale	Nicola Mezucelli	1868				Abruzzo	Teramo	Teramo	Corso San Giorgio
.198		Teatro Comunale	Romano Dal Maso	1905	1970			Veneto	Vicenza	Thiene	Viale Francesco Bassani, 18/22
.199		Teatro Comunale	Carlo Gatteschi	1872				Umbria	Perugia	Todi	Via Mazzini
.200		Teatro Comunale		1872				Sicilia	Catania	Trecastagni	Corso Sicilia, 53

Site of arch. int.	UNESCO Site	ERHT Site	Integrated/Isolated	Construction	Seats	Level	Keyword	Notes	Code
No	No	No	Isolated	Reinforced concrete	379	Regional			.191
Yes	No	No	Isolated	Mixed structure		Local			.192
Yes	No	No	Isolated	Mixed structure	305	Regional			.193
No	No	No	Isolated	Mixed structure	390	Regional		Unusable due to the 2012 earthquake	.194
Yes	No	No	Isolated	Wooden and brick	540	Regional		Closed	.195
Yes	No	No	Isolated	Mixed structure	700	Regional			.196
Yes	No	No	Isolated	Wooden and brick	608	Regional		Demolished in 1959 and replaced with the new theatre	.197
Yes	No	No	Isolated	Mixed structure	500	Regional			.198
Yes	No	No	Isolated	Wooden and brick		Local			.199
Yes	No	No	Isolated	Wooden and brick		Local			.200

Code	Code2	Name	Architect/ Engineer	Year	R1	R2	R3	Region	Province	City	Address
.201		Teatro Comunale	Carlo Rusca	1821	1865	1982		Marche	Macerata	Treia	Piazza P. Arcangeli, 1
.202		Teatro Comunale	Giuseppe Di Bartolo Morselli	1877				Sicilia	Ragusa	Vittoria	Via Garibaldi, 143
.203		Teatro Comunale Alfieri	Raffaele Grilli	1876	1985			Marche	Ancona	Montemarciano	Via Umberto I, 34
.204		Teatro Comunale Camillo De Nardis	Camillo Salvini	1890	1994			Abruzzo	Chieti	Orsogna	Piazza Giuseppe Mazzini, 1
.205		Teatro Comunale Carlo Goldoni	Filippo Antolini	1845	1986			EmiliaRomagna	Ravenna	Bagnocavallo	Piazza Libertà, 18
.206		Teatro Comunale Ciro Pinsuti	Gian Paolo Terrosi	1807	2002			Toscana	Siena	Sinaluga	Via San Martino
.207	*CODE 5.02*	*Teatro Comunale Claudio Abbado*	*Morelli, Foschini*	*1797*	*1949*	*1961*	*1987*	*EmiliaRomagna*	*Ferrara*	*Ferrara*	*Corso Martiri della Libertà, 5*
.208		Teatro Comunale Concordia		1843				EmiliaRomagna	Ferrara	Medolla	Corso Vittorio Emanuele II, 52

(continued)

(continued)

Code	Code2	Name	Architect/ Engineer	Year	R1	R2	R3	Region	Province	City	Address
.209		Teatro Comunale Concordia	Ignazio Cantalamessa	1827	1929	2001		Marche	Ascoli Piceno	San Benedetto del Tronto	Largo Giuseppe Mazzini, 1
.210		Teatro Comunale degli Arrischianti		1740	1985			Toscana	Siena	Sarteano	Piazza XXIV Giugno, 3

List of Historical Theatres in Italy

Site of arch. int.	UNESCO Site	ERHT Site	Integrated/Isolated	Construction	Seats	Level	Keyword	Notes	Code
Yes	No	No	Isolated	Mixed structure	270	Local			.201
Yes	No	No	Isolated	Wooden and brick	380	Regional			.202
Yes	No	No	Isolated	Mixed structure	148	Local			.203
Yes	No	No	Isolated	Mixed structure	240	Local			.204
Yes	No	No	Isolated	Mixed structure	373	Regional			.205
Yes	No	No	Isolated	Mixed structure	150	Local			.206
Yes	*Yes*	*No*	*Isolated*	*Mixed structure*	*900*	*Regional*	*Architectural node*	*Damaged by the 2012 earthquake*	*.207*
Yes	No	No	Isolated	Wooden and brick		Regional			.208
Yes	No	No	Isolated	Mixed structure	312	Regional			.209
Yes	No	No	Isolated	Mixed structure	146	Local			.210

Code	Code2	Name	Architect/Engineer	Year	R1	R2	R3	Region	Province	City	Address
.211		Teatro Comunale degli Oscuri	Carlo Mannucci Benincasa	1870	1982			Toscana	Siena	Torrito di Siena	Piazza Giacomo Matteotti, 10
.212		Teatro Comunale di Laurino		1765	1978			Campania	Salerno	Laurino	Via del teatro
.213		Teatro Comunale Elisabetta Turroni		1867	1994			EmiliaRomagna	Forli-Cesena	Sogliano al Rubicone	Piazza Mazzini, 18
.214		Teatro Comunale Eschilo		1832	2000			Sicilia	Caltanissetta	Gela	Piazza Sant'Agostino
.215		Teatro Comunale ex Accademia dei Georgofili Accalorati		1780	1999			Toscana	Siena	San Casciano dei Bagni	Via Luzia
.216		Teatro Comunale Fedele Fenaroli	Taddeo Salvini	1841	1933	1998		Abruzzo	Chieti	Lanciano	Via dei Frentani, 6
.217		Teatro Comunale Ferdinando Bibiena	Emanuele Brachini	1888	1998			EmiliaRomagna	Bologna	Sant'Agata Bolognese	Via 2 Agosto 1980, 114
.218		Teatro Comunale Ferruccio Martini		1853	1907	1969	2007	Veneto	Rovigo	Trecenta	Piazza Giuseppe Garibaldi
.219		Teatro Comunale Filarmonico	Giuliano Facchinetti	1862				Veneto	Padova	Piove di Sacco	Via Cavour, 19
.220		Teatro Comunale Francesco Cavalli		1893				Piemonte	Alessandria	San Salvatore Monferrato	Via Tarchetti

List of Historical Theatres in Italy

Site of arch. int.	UNESCO Site	ERHT Site	Integrated/Isolated	Construction	Seats	Level	Keyword	Notes	Code
Yes	No	No	Isolated	Mixed structure	120	Local			.211
Yes	No	No	Isolated	Mixed structure		Local			.212
Yes	No	No	Isolated	Mixed structure	120	Local			.213
Yes	No	No	Isolated	Mixed structure	300	Regional			.214
Yes	No	No	Isolated	Mixed structure	90	Local			.215
Yes	No	No	Isolated	Mixed structure	500	Regional			.216
Yes	No	No	Integrated	Mixed structure	120	Local			.217
Yes	No	No	Isolated	Reinforced concrete		Local			.218
Yes	No	No	Integrated	Wooden and brick		Local			.219
Yes	No	No	Isolated	Wooden and brick	500	Regional			.220

Code	Code2	Name	Architect/Engineer	Year	R1	R2	R3	Region	Province	City	Address
.221		Teatro Comunale Francesco Cilèa		1931	1985			Calabria	Reggio Calabria	Reggio Calabria	Corso Garibaldi
.222		Teatro Comunale Gabriello Chiabrera	Carlo Falconieri	1853	1883	1954	2002	Liguria	Savona	Savona	Piazza Armando Diaz, 2
.223		Teatro Comunale Garibaldi		1876				Sicilia	Siracusa	Avola	Via Marconi
.224		Teatro Comunale Giuseppe Manini	Giovanni Santini	1856				Umbria	Terni	Narni	Via Giuseppe Garibaldi
.225		Teatro Comunale La Casa del Popolo		1907	1925	1981		EmiliaRomagna	Bologna	Castello d'Argile	Via Matteotti
.226		Teatro Comunale Malatesta		1868	1994			EmiliaRomagna	Ravenna	Montefiore Conca	Via della Repubblica, 2
.227		Teatro Comunale Mancinelli	Giovanni Santini	1863	1993			Umbria	Terni	Orvieto	Corso Cavour, 122
.228		Teatro Comunale Mario Del Monaco	Miazzi, Scala	1692	1869			Veneto	Treviso	Treviso	Corso del Popolo, 31
.229		Teatro Comunale Masini	Giuseppe Pistocchi	1787	1850	1908	1980	EmiliaRomagna	Ravenna	Faenza	Piazza Nenni, 3
.230		Teatro Comunale Mauro Pagano	Giovanni Battista Vergani	1827				Lombardia	Mantova	Canneto sull'Oglio	Via Mario Corradini, 16

List of Historical Theatres in Italy

Site of arch. int.	UNESCO Site	ERHT Site	Integrated/Isolated	Construction	Seats	Level	Keyword	Notes	Code
Yes	No	No	Isolated	Mixed structure	1200	National			.221
Yes	No	No	Isolated	Mixed structure	626	Regional			.222
Yes	No	No	Isolated	Wooden and brick	346	Regional			.223
Yes	No	No	Isolated	Wooden and brick	350	Regional			.224
No	No	No	Isolated	Mixed structure	156	Local			.225
Yes	No	No	Integrated	Mixed structure	168	Local			.226
Yes	No	No	Isolated	Mixed structure	560	Regional			.227
Yes	No	No	Integrated	Mixed structure	650	Regional			.228
Yes	No	Yes	Integrated	Mixed structure	498	Regional			.229
Yes	No	No	Integrated	Wooden and brick	244	Local			.230

Code	Code2	Name	Architect/Engineer	Year	R1	R2	R3	Region	Province	City	Address
.231		Teatro Comunale Millico	Giovanni Castelli	1878				Puglia	Bari	Terlizzi	Piazza IV Novembre, 19
.232		Teatro Comunale Politeama		1790	1850	1975	1982	EmiliaRomagna	Bologna	San Giovanni in Persiceto	Corso Italia, 72
.233		Teatro Comunale Ponchielli	Luigi Canonica	1806	1989			Lombardia	Cremona	Cremona	Corso Vittorio Emanuele, 52
.234		Teatro Comunale Riccardo Casalaina		1700				Sicilia	Messina	Novara di Sicilia	Via Bellini
.235		Teatro Comunale Rolando Ricci		1911				EmiliaRomagna	Ferrara	Goro	Via Cesare Battisti, 72
.236		Teatro comunale Sollima		1807	1994			Sicilia	Trapani	Marsala	Via Teatro
.237		Teatro Comunale Tina Di Lorenzo		1870	1921	1990		Sicilia	Siracusa	Noto	Piazza 16 Maggio, 1
.238		Teatro Comunale Umberto Giordano		1825	1986			Puglia	Foggia	Foggia	Piazza Cesare Battisti
.239		Teatro Comunale Vincenzo Bellini		1870	1981			Lombardia	Cremona	Casalbuttano ed Uniti	Via Jacini, 23
.240		Teatro Comunale Vittorio Emanuele II	Pasquale Francesconi	1855				Campania	Benevento	Benevento	Corso Giuseppe Garibaldi, 1

List of Historical Theatres in Italy

Site of arch. int.	UNESCO Site	ERHT Site	Integrated/Isolated	Construction	Seats	Level	Keyword	Notes	Code
Yes	No	No	Isolated	Wooden and brick	300	Regional			.231
Yes	No	Yes	Isolated	Mixed structure	235	Local			.232
Yes	No	No	Isolated	Mixed structure	1188	National		Previous theatre of 1747 destroyed by fire	.233
Yes	No	No	Isolated	Wooden and brick		Local			.234
Yes	No	No	Isolated	Reinforced concrete	98	Local			.235
Yes	No	No	Isolated	Mixed structure	300	Regional			.236
Yes	No	No	Isolated	Mixed structure	320	Local			.237
Yes	No	No	Isolated	Mixed structure	470	Regional			.238
Yes	No	No	Isolated	Mixed structure	289	Local			.239
Yes	No	No	Isolated	Wooden and brick	500	Regional			.240

Code	Code2	Name	Architect/Engineer	Year	R1	R2	R3	Region	Province	City	Address
.241		Teatro Comunale Wanda Capodaglio		1930	1969	1987		Toscana	Arezzo	Castelfranco Piandiscò	Via Roma, 34
.242		Teatro Comunale Weber Facchini	Araldo Vincenzi	1927				EmiliaRomagna	Modena	Portomaggiore	Piazza del Popolo
.243		Teatro Condominale la Fortuna		1758	1926	1966		Marche	Ancona	Monte San Vito	Via Guglielmo Marconi, 5
.244		Teatro Consorziale		1672	1924	1986	2003	EmiliaRomagna	Bologna	Budrio	Via G. Garibaldi, 35
.245		Teatro Contavalli	Martinetti, Nadi	1814	1923			EmiliaRomagna	Bologna	Bologna	Via Guglielmo Oberdan
.246		Teatro Conti	Marco Severini	1904	1968			Marche	Pesaro-Urbino	Acqualagna	Via Enrico Mattei, 12/13
.247		Teatro Continuo	Alberto Burri	1973				Lombardia	Milano	Milano	Parco Sempione
.248		Teatro Convitto Nazionale Cicognini		1783				Toscana	Prato	Prato	Piazza del Collegio, 13
.249		Teatro Coppola		1821				Sicilia	Catania	Catania	Via del Vecchio Bastione, 9
.250		Teatro Cortesi	Achille Buffoni	1873	1989			Marche	Ancona	Sirolo	Via del Teatro, 1/3

List of Historical Theatres in Italy

Site of arch. int.	UNESCO Site	ERHT Site	Integrated/Isolated	Construction	Seats	Level	Keyword	Notes	Code
No	No	No	Isolated	Reinforced concrete	300	Regional			.241
Yes	No	No	Isolated	Mixed structure	252	Local			.242
Yes	No	No	Isolated	Mixed structure	150	Local			.243
Yes	No	No	Integrated	Mixed structure	550	Regional			.244
Yes	No	No	Isolated	Mixed structure	800	Regional			.245
No	No	No	Isolated	Mixed structure	320	Regional			.246
Yes	No	Yes	Isolated	Mixed structure		International		Temporary, demolished in 1989, rebuilt for EXPO 15	.247
Yes	No	No	Integrated	Wooden and brick	700	Regional		Destroyed by fire in 1943, self-managed since 2011	.248
Yes	No	No	Isolated	Wooden and brick		Regional			.249
Yes	No	No	Isolated	Mixed structure	220	Local			.250

Code	Code2	Name	Architect/Engineer	Year	R1	R2	R3	Region	Province	City	Address
.251		Teatro Curci		1877				Puglia	Barletta-Andria-Trani	Barletta	Corso Vittorio Emanuele, 71
.252		Teatro Dal Verme	Giuseppe Pestagalli	1872	1990			Lombardia	Milano	Milano	Via San Giovanni Sul Muro, 2
.253		Teatro d'Angennes		1780				Piemonte	Torino	Torino	Via Principe Amedeo, 26
.254		Teatro Dante	Mariano Falcini	1873	2007			Toscana	Firenze	Campi Bisenzio	Piazza Dante Alighieri, 23
.255		*Teatro De La Sena*	*Giannantonio Selva*	*1684*	*1810*			*Veneto*	*Belluno*	*Feltre*	*Piazza Maggiore*
.256		Teatro De Larderel		1872				Toscana	Pisa	Pomarance	Via Roncalli
.257		Teatro De Micheli		1909	1923	2006		EmiliaRomagna	Ferrara	Copparo	Piazza del Popolo, 11/A
.258		Teatro degli Animosi		1792	1990			Toscana	Firenze	Marradi	Viale della Repubblica, 3
.259		Teatro degli Animosi	Giuseppe Pardini	1840				Toscana	Massa-Carrara	Massa Carrara	Piazza Cesare Battisti, 1
.260		Teatro degli Antei		1726	1987			Toscana	Arezzo	Pratovecchio	Via Giuseppe Verdi

List of Historical Theatres in Italy

Site of arch. int.	UNESCO Site	ERHT Site	Integrated/Isolated	Construction	Seats	Level	Keyword	Notes	Code
Yes	No	No	Isolated	Wooden and brick	170	Local			.251
Yes	No	Yes	Isolated	Mixed structure	1500	National		Destroyed by bombing in 1943 and rebuilt	.252
Yes	No	No	Isolated	Wooden and brick	1300	National		Turned into a cinema	.253
Yes	No	No	Isolated	Mixed structure		Local			.254
Yes	*No*	*No*	*Integrated*	*Wooden and brick*	*298*	*Regional*	*Knotting*		.255
Yes	No	No	Isolated	Wooden and brick	98	Local			.256
Yes	No	No	Isolated	Mixed structure	496	Regional			.257
Yes	No	No	Isolated	Mixed structure	257	Local			.258
Yes	No	No	Isolated	Wooden and brick	440	Regional		Currently closed	.259
Yes	No	No	Isolated	Mixed structure	310	Regional			.260

Code	Code2	Name	Architect/ Engineer	Year	R1	R2	R3	Region	Province	City	Address
.261		Teatro degli Astrusi	Leonardo De Vegni	1766	1992			Toscana	Siena	Montalcino	Via Bandi, 1
.262		Teatro degli Avvaloranti		1834	2003			Umbria	Perugia	città della Pieve	Via Pò di Mezzo
.263		Teatro degli Illuminati	Tifernate Antonio Gabrielli	1666	1920			Umbria	Perugia	Città di Castello	Via dei Fucci
.264		Teatro degli Industri		1819	1888	1989		Toscana	Grosseto	Grosseto	Via Mazzini, 99
.265		Teatro degli Unanimi		1741				Toscana	Grosseto	Arcidosso	Piazza Cavallotti, 4
.266		*Teatro dei Cavalieri Associati*	*Cosimo Morelli*	*1782*				*EmiliaRomagna*	*Forlì-Cesena*	*Imola*	*Vie Emilia, 80*
.267		Teatro dei Concordi	Francesco Fedi	1867	1964	1990		Toscana	Livorno	Campiglia Marittima	Via Aldo Moro, 1
.268		Teatro dei Concordi		1770				Toscana	Firenze	Firenze	Corso dei Tintori, 35
.269		Teatro dei Concordi		1909	1980			Toscana	Siena	Montepulciano	Via Fratelli Braschi
.270		Teatro dei Concordi	Timoleone Crocchi	1872	1923	1970	1991	Toscana	Grosseto	Roccastrada	Corso Roma, 8

List of Historical Theatres in Italy

Site of arch. int.	UNESCO Site	ERHT Site	Integrated/Isolated	Construction	Seats	Level	Keyword	Notes	Code
Yes	No	No	Isolated	Mixed structure	180	Local			.261
Yes	No	No	Isolated	Mixed structure	200	Local			.262
Yes	No	No	Isolated	Mixed structure		National			.263
Yes	No	No	Isolated	Mixed structure	350	Regional			.264
Yes	No	No	Isolated	Wooden and brick	120	Local			.265
Yes	*No*	*No*	*Integrated*	*Wooden and brick*		*Regional*		*Destroyed by fire in 1797*	.266
Yes	No	No	Isolated	Mixed structure	190	Local			.267
Yes	Yes	No	Isolated	Wooden and brick		Regional		Closed in 1954	.268
Yes	No	No	Isolated	Mixed structure	200	Local			.269
Yes	No	No	Isolated	Mixed structure	190	Local			.270

Code	Code2	Name	Architect/Engineer	Year	R1	R2	R3	Region	Province	City	Address
.271		Teatro dei Coraggiosi	Ferdinando Magagnini	1862	1985			Toscana	Pisa	Pomarance	Via Gramsci, 15/A
.272		Teatro dei Differenti		1792	1985			Toscana	Lucca	Barga	Via di Mezzo, 45
.273		Teatro dei Dovizi	Niccolò Matas	1842	1997			Toscana	Arezzo	Bibbiena	Via Rosa Scoti Franceschi, 21
.274		Teatro dei Filarmonici		1831	1994			Marche	Ascoli Piceno		Via delle Torri
.275		Teatro dei Filodrammatici		1909	2000			EmiliaRomagna	Piacenza	Piacenza	Via Santa Franca, 33
.276		Teatro dei Leggieri	Francesco Marinelli	1794	1997			Toscana	Siena	San Gimignano	Piazza del Duomo
.277	*CODE 5.05*	*Teatro dei Principi di Carignano*	*Benedetto Alfieri*	*1753*	*1935*	*2007*		*Piemonte*	*Torino*	*Torino*	*Piazza Carignano, 6*
.278		Teatro dei Pupazzi Borgobello	Belardi, Bori, Burini	2010				Umbria	Perugia	Perugia	Via del Castellano
.279		Teatro dei Rassicurati	Antonio Capretti	1795	1894	1973	1994	Toscana	Lucca	Montecarlo	Via Carmignani, 14
.280		Teatro dei Rinnovati		1800				Veneto	Treviso	Asolo	Piazzetta San Pio X, 192

List of Historical Theatres in Italy

Site of arch. int.	UNESCO Site	ERHT Site	Integrated/Isolated	Construction	Seats	Level	Keyword	Notes	Code
Yes	No	No	Isolated	Mixed structure	220	Local			.271
Yes	No	No	Isolated	Mixed structure	286	Local			.272
Yes	No	No	Isolated	Mixed structure	100	Local			.273
Yes	No	No	Isolated	Mixed structure	400	Regional			.274
Yes	No	No	Isolated	Mixed structure	300	Regional			.275
Yes	No	No	Isolated	Mixed structure		Regional			.276
Yes	*No*	*No*	*Isolated*	*Mixed structure*	*875*	*National*	*Architectural node*		.277
No	No	No	Isolated	Wooden and brick	94	Local			.278
Yes	No	No	Isolated	Mixed structure	170	Local		Previous theatre of 1647 designed by Carlo Fontana destroyed by fire	.279
Yes	No	No	Integrated	Wooden and brick		Regional			.280

Code	Code2	Name	Architect/Engineer	Year	R1	R2	R3	Region	Province	City	Address
281		*Teatro dei Rinnovati*	*Antonio Galli Bibiena*	*1750*				*Toscana*	*Siena*	*Siena*	*Piazza del Campo, 1*
282		Teatro dei Risorti		1820				Toscana	Pistoia	Montecatini Alto	Piazza Giuseppe Giusti
283		Teatro dei Riuniti	Giovanni Cerrini	1813				Umbria	Perugia	Umbertide	Piazza Fortebraccio, 1
284		Teatro dei Rozzi	Alessandro Doveri	1817	1998			Toscana	Siena	Siena	Piazza dell'Indipendenza, 15
285		Teatro dei Rustici		1732	1990			Umbria	Terni	Monteleone d'Orvieto	Piazza del Municipio, 3
286		Teatro dei Varii	Antonio Galli Bibiena	1762	1982			Toscana	Siena	Colle di Val d'Elsa	Via del Castello, 64
287		Teatro dei Vigilanti	Paolo Bargigli	1815	1922	1937	1980	Toscana	Siena	Portoferraio	Piazza Gramsci
288		Teatro del Carmine		1929				Sardegna	Sassari	Tempio Pausania	Piazza del Carmine, 6
289		Teatro del Collegio di San Carlo		1753	1929	1989		EmiliaRomagna	Modena	Modena	Via San Carlo, 5
290		Teatro del Convitto Nazionale Maria Luigia		1829	2017			EmiliaRomagna	Parma	Parma	Borgo Lalatta, 14

List of Historical Theatres in Italy

Site of arch. int.	UNESCO Site	ERHT Site	Integrated/Isolated	Construction	Seats	Level	Keyword	Notes	Code
Yes	*Yes*	*No*	*Integrated*	*Wooden and brick*	*549*	*Regional*	*Knotting*		*.281*
Yes	No	No	Isolated	Wooden and brick		Local			.282
Yes	No	No	Isolated	Wooden and brick		Local			.283
Yes	Yes	No	Isolated	Mixed structure	499	Regional			.284
Yes	No	No	Isolated	Mixed structure	96	Local			.285
Yes	No	No	Isolated	Mixed structure	200	Local			.286
Yes	No	No	Isolated	Mixed structure	225	Local			.287
Yes	No	No	Isolated	Reinforced concrete	100	Local			.288
Yes	No	No	Integrated	Mixed structure	180	Local			.289
Yes	No	No	Integrated	Mixed structure	178	Local			.290

Code	Code2	Name	Architect/Engineer	Year	R1	R2	R3	Region	Province	City	Address
.291		Teatro del Genio		1721	1924			Lazio	Viterbo	Viterbo	Via Teatro del Genio, 16
.292		Teatro del Giglio		1817	1911	1957	1983	Toscana	Lucca	Lucca	Piazza del Giglio, 13/15
.293		Teatro del Leone	Pietro Maggi	1815	1997			Marche	Fermo	Santa Vittoria in Materano	Corso Giacomo Matteotti, 13
.294	CODE 1.01	Teatro del Mondo	Aldo Rossi	1979				Veneto	Venezia	Venezia	
.295		Teatro del Palazzo di Bassano		1790				Lazio	Viterbo	Bassano Romano	Piazza Umberto I
.296		Teatro del Pavone	Pietro Carattoli	1717				Umbria	Perugia	Perugia	Piazza Repubblica
.297		Teatro del Popolo	Giuseppe Niccoli	1867				Toscana	Firenze	Castelfiorentino	Piazza Gramsci, 77
.298		Teatro del Popolo		1921				Lombardia	Milano	Gallarate	Via Palestro, 5
.299		Teatro del Popolo	Bettino Marchetti	1890	1983	2001		Toscana	Siena	Rapolano Terme	Piazzale del Teatro
.300		Teatro del Serpente Aureo	Pietro Maggi	1862				Marche	Ascoli Piceno	Offida	Piazza del Popolo

List of Historical Theatres in Italy

Site of arch. int.	UNESCO Site	ERHT Site	Integrated/Isolated	Construction	Seats	Level	Keyword	Notes	Code
Yes	No	No	Isolated	Mixed structure		Local		Destroyed by bombing in 1943 and rebuilt in 1948	.291
Yes	No	No	Isolated	Mixed structure	749	Regional			.292
Yes	No	No	Isolated	Mixed structure	144	Local			.293
Yes	*Yes*	*No*	*Isolated*	*Wooden and still*		*International*	*Enclosure*	*Temporary*	*.294*
Yes	No	No	Integrated	Wooden		Local			.295
Yes	No	No	Isolated	Wooden and brick		Regional			.296
Yes	No	No	Isolated	Wooden and brick	148	Local			.297
Yes	No	No	Isolated	Mixed structure	200	Local			.298
Yes	No	No	Isolated	Mixed structure	250	Local			.299
Yes	No	No	Isolated	Wooden and brick	500	Regional			.300

Code	Code2	Name	Architect/Engineer	Year	R1	R2	R3	Region	Province	City	Address
.301		Teatro del Trionfo		1801	1866	1919	2015	Marche	Pesaro-Urbino	Cartoceto	Piazzale Guglielmo Marconi
.302	*CODE 4.01 and 4.21*	*Teatro della Compagnia*	*Adolfo Natalini*	*1987*				*Toscana*	*Firenze*	*Firenze*	*Via Camillo Cavour, 50/R*
.303		Teatro della Concordia		1873	2012			Umbria	Perugia	Marsciano	Largo Goldoni, 9
.304		Teatro della Concordia		1808	1993			Umbria	Perugia	Monte Castello di Vibio	Piazza del Teatro, 4
.305		Teatro della Concordia		1843	2010			Sicilia	Ragusa	Ragusa	Via Ecce Homo, 188
.306		Teatro della Concordia		1750	1935	1987		Marche	Pesaro-Urbino	San Costanzo	Via Montegrappa, 4
.307		Teatro della Filarmonica		1899				Umbria	Perugia	Corciano	Via del Serraglio
.308		*Teatro della Fortuna*	*Torelli; Poletti*	*1677*	*1863*	*1998*		*Marche*	*Pesaro-Urbino*	*Fano*	*Piazza XX Settembre, 1*
.309		Teatro della Grancia di Montisi		1899	2005			Toscana	Siena	Montisi	Via Umberto I, 8
.310		Teatro della Misericordia		1923	2005			Toscana	Firenze	Vinci	Via Pierino Da Vinci, 39

List of Historical Theatres in Italy

Site of arch. int.	UNESCO Site	ERHT Site	Integrated/Isolated	Construction	Seats	Level	Keyword	Notes	Code
Yes	No	No	Isolated	Mixed structure	200	Local			.301
Yes	*Yes*	*No*	*Integrated*	*Reinforced concrete*	*500*	*Regional*	*Recast*		*.302*
Yes	No	No	Isolated	Mixed structure		Local			.303
Yes	No	No	Isolated	Mixed structure	99	Local			.304
Yes	No	No	Isolated	Mixed structure	500	Regional			.305
Yes	No	No	Isolated	Mixed structure	150	Local			.306
Yes	No	No	Integrated	Wooden and brick	144	Local			.307
Yes	*No*	*No*	*Integrated*	*Mixed structure*	*595*	*Regional*	*Knotting*	*Damaged during the second world war*	*.308*
Yes	No	No	Integrated	Mixed structure	58	Local			.309
Yes	No	No	Isolated	Reinforced concrete		Local			.310

Code	Code2	Name	Architect/Engineer	Year	R1	R2	R3	Region	Province	City	Address
.311	*CODE 5.03*	*Teatro della Pergola*	*Ferdinando Tacca*	*1652*	*1948*	*1967*	*1988*	*Toscana*	*Firenze*	*Firenze*	*Via della Pergola, 18*
.312		Teatro della Rocca		1860				Marche	Pesaro-Urbino	Sassocorvaro	Piazza Battelli
.313		Teatro della Rocca Franco Tagliavini		1868	1984			EmiliaRomagna	Reggio Emilia	Novellara	Piazzale Marconi, 1
.314		Teatro della Sala		1636	1930			EmiliaRomagna	Bologna	Bologna	
.315		Teatro della Sapienza		1829	1999			Umbria	Perugia	Perugia	Via della Sapienza, 6
.316		Teatro della Società	Giuseppe Bovara	1844	1979	1994		Lombardia	Lecco	Lecco	Piazza Garibaldi, 10
.317		Teatro della Società dei Filodrammatici		1810				Marche	Macerata	Macerata	Via Antonio Gramsci, 30
.318		Teatro della Vittoria	Luigi Fedeli	1834	1935	1947	1991	Marche	Macerata	Sarnano	Piazza Alta
.319		Teatro dell'Accademia	Andrea Scala	1869				Veneto	Treviso	Conegliano	Piazza Cima, 5
.320		Teatro dell'Accademia		1899	2004			Umbria	Perugia	Tuoro	Via della Croce, 1

List of Historical Theatres in Italy

Site of arch. int.	UNESCO Site	ERHT Site	Integrated/Isolated	Construction	Seats	Level	Keyword	Notes	Code
Yes	*Yes*	*No*	*Isolated*	*Mixed structure*	*999*	*National*	*Knotting*		*.311*
Yes	No	No	Integrated	Wooden and brick	100	Local			.312
Yes	No	No	Isolated	Wooden and brick	399	Regional			.313
Yes	No	No	Integrated	Wooden and brick		Regional		Different use	.314
Yes	No	No	Isolated	Wooden and brick	127	Local			.315
Yes	No	No	Isolated	Wooden and brick	464	Regional			.316
Yes	No	No	Isolated	Wooden and brick		Regional			.317
Yes	No	No	Isolated	Mixed structure	150	Local			.318
Yes	No	No	Integrated	Wooden and brick	803	Regional			.319
Yes	No	No	Isolated	Mixed structure	178	Local			.320

Code	Code2	Name	Architect/Engineer	Year	R1	R2	R3	Region	Province	City	Address
.321		Teatro dell'Accademia dei Ricomposti	Lorenzo Pozzolini	1789	1936	1962		Toscana	Arezzo	Anghiari	Via Bozia, 3
.322		Teatro dell'Accademia della Rosa		1753	1840	1948	1998	Toscana	Massa-Carrara	Pontremoli	Piazza del Teatro
.323		Teatro dell'Affratellamento		1876	2003			Toscana	Firenze	Firenze	Via Giampaolo Orsini, 73
.324		Teatro dell'Aquila	Cosimo Morelli	1790				Marche	Fermo	Fermo	Via Giuseppe Mazzini, 4
.325		Teatro dell'Arancio	Pietro Maggi	1790	2003			Marche	Fermo	Grottammare	Via C. Peretti
.326		Teatro delle Commedie	Gherardesca, Della Valle	1864	1979	2010		Toscana	Livorno	Livorno	Via Giuseppe Maria Terreni, 5
.327		Teatro delle Dame	Francesco Galli Bibiena	1718				Lazio	Roma	Roma	Via Margutta
.328		Teatro delle Logge	Pietro Augustoni	1808	2003			Marche	Macerata	Montecosaro	Piazza Trieste
.329		*Teatro delle Muse*	*Ghinelli, Guerri, Salmoni*	*1827*	*2002*			*Marche*	*Ancona*	*Ancona*	*Piazza della Repubblica*
.330		Teatro delle Saline	Manovella	1931	1991			Sardegna	Cagliari	Cagliari	Piazzetta Billy Sechi, 3/4

List of Historical Theatres in Italy

Site of arch. int.	UNESCO Site	ERHT Site	Integrated/Isolated	Construction	Seats	Level	Keyword	Notes	Code
Yes	No	No	Isolated	Mixed structure	220	Local			.321
Yes	No	No	Isolated	Mixed structure	253	Local			.322
Yes	Yes	No	Isolated	Mixed structure	126	Local			.323
Yes	No	No	Isolated	Wooden and brick	1000	National			.324
Yes	No	No	Isolated	Mixed structure	250	Local			.325
Yes	No	No	Isolated	Mixed structure	150	Local			.326
Yes	Yes	No	Isolated	Wooden and brick	1400	National		Destroyed by fire in 1868	.327
Yes	No	No	Isolated	Mixed structure	98	Local			.328
Yes	*No*	*No*	*Isolated*	*Mixed structure*	*1147*	*National*	*Architectural node*	*Rebuilt after the bombings of 1943*	*.329*
Yes	No	No	Isolated	Reinforced concrete	332	Regional			.330

Code	Code2	Name	Architect/Engineer	Year	R1	R2	R3	Region	Province	City	Address
.331		Teatro dell'Iride	Giuseppe Sabbatini	1877	1982			Marche	Fermo	Petritoli	Via Teatro, 12
.332		Teatro dell'Istituto Francese		1913				Toscana	Firenze	Firenze	Piazza Ognissanti, 2
.333		Teatro dell'Olivo		1770	1920	2003		Toscana	Lucca	Camaiore	Via Vittorio Emanuele
.334		*Teatro dell'Opera*	*Sfondrini, Piacentini*	*1880*	*1926*	*1960*		*Lazio*	*Roma*	*Roma*	*Piazza Beniamino Gigli, 1*
.335		Teatro dell'Opera del Casinò		1905				Liguria	Imperia	Sanremo	Corso degli Inglesi, 18
.336		Teatro dell'Unione Alagnese		1900				Piemonte	Vercelli	Alagna Valsesia	Località Pedelegno
.337		Teatro di Cestello		1903	1985			Toscana	Firenze	Firenze	Piazza del Cestello, 4
.338	*CODE 5.04*	*Teatro di Corte*	*Luigi Vanvitelli*	*1769*	*1989*			*Campania*	*Caserta*	*Caserta*	*Reggia di Caserta*
.339		Teatro di Palazzo Rinucci		1753				Toscana	Firenze	Firenze	Via Santo Spirito, 39–41
.340	*CODE 2.04*	*Teatro di Palazzo Vecchio*	*Giorgio Vasari*	*1565*				*Toscana*	*Firenze*	*Firenze*	

List of Historical Theatres in Italy

Site of arch. int.	UNESCO Site	ERHT Site	Integrated/Isolated	Construction	Seats	Level	Keyword	Notes	Code
Yes	No	No	Isolated	Mixed structure	220	Local			.331
Yes	Yes	No	Integrated	Reinforced concrete	99	Local			.332
Yes	No	No	Isolated	Mixed structure	255	Local			.333
Yes	Yes	No	Isolated	Mixed structure	1400	International	Architectural node		.334
Yes	No	No	Isolated	Reinforced concrete	400	Regional			.335
No	No	No	Isolated	Wooden and brick		Local			.336
Yes	Yes	No	Isolated	Mixed structure	140	Local			.337
Yes	Yes	No	Integrated	Mixed structure		National	Architectural node		.338
Yes	Yes	No	Isolated	Mixed structure		National			.339
Yes	Yes	No	Integrated	Wooden and brick		National	Organism		.340

Code	Code2	Name	Architect/Engineer	Year	R1	R2	R3	Region	Province	City	Address
.341		Teatro di Rifredi		1914	1980			Toscana	Firenze	Firenze	Via Vittorio Emanuele II, 303
.342		Teatro di San Pietro		1850				Toscana	Pisa	Volterra	Via Don Minzoni, 49
.343		Teatro di Santa Cecilia		1693	1816	1853	2009	Sicilia	Palermo	Palermo	Via Piccola del Teatro S. Cecilia, 5
.344		Teatro di Santa Chiara		1900				Toscana	Pisa	Volterra	Borgo Santo Stefano, 153
.345		Teatro di Torino	Giuseppe Bollati	1857				Piemonte	Torino	Torino	Via Verdi, 26
.346		Teatro di Via Verdi		1900				Toscana	Pisa	Vicopisano	Via Giuseppe Verdi, 12
.347		Teatro di Villa Duchessa di Galleria	Gaetano Cantoni	1780	2010			Liguria	Genova	Genova	Via Nicolò da Corte, 2
.348		Teatro di Villa Patrizi	Ferdinando Sanfelice	1770	2011			Campania	Napoli	Napoli	Via Manzoni, 41
.349		Teatro di Villa Petitot	Ennemond A. Petitot	1770				EmiliaRomagna	Parma	Parma	Strada Madonnina Gigli, 25
.350		Teatro di Villa Raggio		1890				EmiliaRomagna	Piacenza	Pontenure	Parco di Villa Raggio

List of Historical Theatres in Italy

Site of arch. int.	UNESCO Site	ERHT Site	Integrated/Isolated	Construction	Seats	Level	Keyword	Notes	Code
Yes	No	No	Isolated	Mixed structure	286	Local			.341
Yes	No	No	Integrated	Wooden and brick	100	Local			.342
Yes	No	No	Isolated	Mixed structure	336	Regional			.343
Yes	No	No	Integrated	Wooden and brick	90	Local			.344
Yes	No	No	Isolated	Wooden and brick	1400	National		Destroyed by the bombings of 1943	.345
No	No	No	Isolated	Wooden and brick	100	Local			.346
Yes	No	No	Integrated	Mixed structure	100	Local			.347
Yes	Yes	No	Integrated	Mixed structure		Regional			.348
Yes	No	No	Integrated	Wooden and brick		Local		Inside Villa Petitot	.349
Yes	No	No	Integrated	Wooden and brick		Regional		Greenhouse theatre in the villa	.350

Code	Code2	Name	Architect/Engineer	Year	R1	R2	R3	Region	Province	City	Address
.351		Teatro di Villa Torlonia	Quintiliano Raimondi	1874	2013			Lazio	Roma	Roma	Via Nomentana, 70
.352		Teatro Domenico Cimarosa		1889	1929			Campania	Caserta	Aversa	Vicolo del Teatro, 3
.353		Teatro Donnafugata		1850	1997			Sicilia	Ragusa	Ragusa	Via Pietro Novelli, 5
.354		Teatro Duse	Antonio Locatelli	1798	1857	1927	1990	Veneto	Treviso	Asolo	Via Regina Cornaro, 3
.355		Teatro Duse	Antonio Brunetti	1822	1904	2003		EmiliaRomagna	Bologna	Bologna	Via Cartolerie, 42
.356		Teatro Duse	Faustino Colombi	1827	1975	1995		EmiliaRomagna	Piacenza	Cortemaggiore	Via XX Settembre
.357		Teatro Eden	Alfonso Modenesi	1911				Veneto	Treviso	Treviso	Via Daniele Monterumici
.358		Teatro Eliseo		1900				Lazio	Roma	Roma	Via Nazionale, 183
.359		Teatro Ermete Novelli		1914				Marche	Fermo	Grottazzolina	Parco della Rimembranza, 1
.360		Teatro Ermete Novelli		1935				EmiliaRomagna	Rimini	Rimini	Via Alberto Bianchi, 3

List of Historical Theatres in Italy

Site of arch. int.	UNESCO Site	ERHT Site	Integrated/Isolated	Construction	Seats	Level	Keyword	Notes	Code
Yes	Yes	No	Integrated	Mixed structure		Regional			.351
Yes	No	No	Integrated	Mixed structure		Local			.352
Yes	No	No	Integrated	Mixed structure	100	Local		Inside Palazzo Donnafugata	.353
Yes	No	No	Isolated	Mixed structure	180	Local			.354
Yes	No	No	Isolated	Mixed structure	949	National			.355
Yes	No	No	Isolated	Mixed structure		Regional			.356
Yes	No	No	Isolated	Mixed structure	451	Regional			.357
Yes	Yes	No	Integrated	Mixed structure	760	Regional			.358
No	No	No	Isolated	Mixed structure	231	Local			.359
Yes	No	No	Isolated	Reinforced concrete	639	Regional			.360

Code	Code2	Name	Architect/Engineer	Year	R1	R2	R3	Region	Province	City	Address
.361		Teatro Ernesto Rossi		1770	1978			Toscana	Pisa	Pisa	Via del Collegio Ricci
.362		Teatro Errico Petrella	Giulio Turchi	1870				EmiliaRomagna	Forlì-Cesena	Longiano	Via IV Novembre
.363		Teatro Ettore Thesorieri		1767	2014			Umbria	Perugia	Cannara	Piazza Corte Vecchia, 23
.364		Teatro Falcone	Carlo Fontana	1679	2004			Liguria	Genova	Genova	Via Balbi, 10
.365		Teatro Faraggiana		1905				Piemonte	Novara	Novara	Via dei Caccia, 1F
.366	*CODE 3.03 and 3.24*	*Teatro Farnese*	*Giovanni Battista Aleotti*	*1628*	*1867*	*1980*		*EmiliaRomagna*	*Parma*	*Parma*	*Piazza della Pilotta, 9A-13/A*
.367		*Teatro Fedeli*	*Fabrizio Carini Motta*	*1669*				*Lombardia*	*Mantova*	*Mantova*	*Via Cardone*
.368		Teatro Federico e Taddeo Zuccari	Bartolomeo Breccioli	1618	1761	1841	1932	Marche	Pesaro-Urbino	Sant'Angelo in Vado	Piazza Umberto I
.369		Teatro Feronia	Ireneo Aleandri	1832	1985			Marche	Macerata	San Severino Marche	Piazza Del Popolo, 15
.370		Teatro Ferrini		1910				Veneto	Treviso	Adria	Piazza Casellati, 3

List of Historical Theatres in Italy

Site of arch. int.	UNESCO Site	ERHT Site	Integrated/Isolated	Construction	Seats	Level	Keyword	Notes	Code
Yes	No	No	Isolated	Mixed structure		Regional			.361
Yes	No	No	Isolated	Wooden and brick	250	Local			.362
Yes	No	No	Isolated	Mixed structure	233	Local			.363
Yes	No	No	Integrated	Mixed structure		Regional		Destroyed by fire in 1944	.364
Yes	No	No	Isolated	Mixed structure	378	Local			.365
Yes	No	Yes	Integrated	Wooden and brick	3000	International	Organism		.366
Yes	Yes	No	Integrated	Wooden and brick		Regional	Recast		.367
Yes	No	No	Isolated	Mixed structure	227	Local			.368
Yes	No	No	Isolated	Mixed structure	442	Regional			.369
Yes	No	No	Isolated	Mixed structure		Local			.370

Code	Code2	Name	Architect/Engineer	Year	R1	R2	R3	Region	Province	City	Address
.371		Teatro Filarmonica di Verona	Francesco Galli Bibiena	1716				Veneto	Verona	Verona	Via dei Mutilati, 4
.372		Teatro Filodrammatici	Canonica, Caccia Dominioni	1800	1904	1923	1970	Lombardia	Milano	Milano	Via Filodrammatici, 1
.373		Teatro Filodrammatici	Carlo Bedolini	1905	1987			Lombardia	Bergamo	Treviglio	Piazzale del Santuario, 1
.374		Teatro Flavio Vespasiano	Ghinelli, Sfondrini	1893	1990	2005		Lazio	Rieti	Rieti	Via Garibaldi
.375		Teatro Flora	Antonio Liozzi	1780	1985			Marche	Macerata	Penna San Giovanni	Piazza del Teatro, 1
.376		Teatro Francesco di Bartolo		1842	1987			Toscana	Pisa	Buti	Via F.lli Disperati, 10
.377		Teatro Francesco Paolo Tosti		1910	1996			Abruzzo	Chieti	Ortona	Corso Garibaldi, 7
.378		Teatro Francesco Stabile	Alvino, Pisanti	1881	1980			Basilicata	Potenza	Potenza	Piazza Mario Pagano
.379		Teatro Fraschini	Antonio Galli Bibiena	1773	1904	1994		Lombardia	Pavia	Pavia	Corso Strada Nuova, 138
.380		Teatro Fuori Squadro		1930	2007			Calabria	Reggio Calabria	Caulonia	Via Ilario Antonio De Blasi

Site of arch. int.	UNESCO Site	ERHT Site	Integrated/Isolated	Construction	Seats	Level	Keyword	Notes	Code
Yes	Yes	No	Isolated	Wooden and brick	1200	Local		Destroyed by fire in 1749, destroyed by bombing in 1945, rebuilt	.371
Yes	*No*	*No*	*Isolated*	*Mixed structure*	*1000*	*National*	*Architectural node*	*Liberty facade of 1904*	*.372*
Yes	No	No	Isolated	Mixed structure	290	Local			.373
Yes	No	No	Isolated	Mixed structure	550	Regional			.374
Yes	No	No	Isolated	Mixed structure	99	Local			.375
Yes	No	No	Isolated	Mixed structure	220	Local			.376
Yes	No	No	Isolated	Mixed structure	400	Regional			.377
Yes	No	No	Isolated	Mixed structure	361	Regional			.378
Yes	No	No	Isolated	Mixed structure	409	Regional			.379
No	No	No	Isolated	Mixed structure		Local			.380

Code	Code2	Name	Architect/Engineer	Year	R1	R2	R3	Region	Province	City	Address
.381		Teatro Garibaldi	Giuseppe Albrizio	1872	2003			Puglia	Barletta-Andria-Trani	Bisceglie	Piazza Margherita di Savoia
.382		Teatro Garibaldi		1932	1984	2001		Sicilia	Enna	Enna	Piazza Umberto I
.383		Teatro Garibaldi	Angelo Pierallini	1872	1983			Toscana	Firenze	Figline Valdarno	Piazza Serristori
.384		Teatro Garibaldi	Leonardo De Vegni	1784				Toscana	Arezzo	Foiano della Chiana	Via Ricasoli, 19–25
.385		Teatro Garibaldi		1825	1879	2015		Puglia	Lecce	Gallipoli	Via Giuseppe Garibaldi, 8
.386		Teatro Garibaldi		1837	2005			Puglia	Foggia	Lucera	Corso Giuseppe Garibaldi, 74
.387		Teatro Garibaldi	Gaspare Viviani	1848				Sicilia	Trapani	Mazara del Vallo	Via del Carmine
.388		Teatro Garibaldi		1820	1857	2000		Sicilia	Ragusa	Modica	Corso Umberto I, 207
.389		Teatro Garibaldi	Pietro Cutrera	1861	2010			Sicilia	Palermo	Palermo	Via Castrofilippo, 38–30
.390		Teatro Garibaldi	Antonio Curri	1890	2002			Campania	Caserta	Santa Maria Capua Vetere	Corso Garibaldi, 78

List of Historical Theatres in Italy

Site of arch. int.	UNESCO Site	ERHT Site	Integrated/Isolated	Construction	Seats	Level	Keyword	Notes	Code
Yes	No	No	Isolated	Mixed structure	400	Regional			.381
Yes	No	No	Isolated	Reinforced concrete		Regional		Reconstruction of the previous theatre of 1872	.382
Yes	No	No	Isolated	Mixed structure	490	Regional			.383
Yes	No	No	Isolated	Wooden and brick		Local			.384
Yes	No	No	Isolated	Mixed structure	150	Local			.385
Yes	No	No	Isolated	Mixed structure	100	Local			.386
Yes	No	No	Integrated	Wooden and brick	99	Local			.387
Yes	No	No	Isolated	Mixed structure	320	Regional			.388
Yes	No	No	Isolated	Mixed structure	150	Local			.389
Yes	No	No	Isolated	Mixed structure	378	Regional			.390

Code	Code2	Name	Architect/Engineer	Year	R1	R2	R3	Region	Province	City	Address
.391		Teatro Gentile da Fabriano	Rossi, Petrini	1884				Marche	Ancona	Fabriano	Via Gentile da Fabriano, 1
.392		Teatro Gerolamo		1868	2017			Lombardia	Milano	Milano	Piazza Cesare Beccaria, 8
.393		Teatro Giacomo Leopardi		1877	2005			Marche	Macerata	San Ginesio	Piazza A. Gentile
.394		Teatro Giacosa	Maurizio Storero	1834	1998			Piemonte	Torino	Ivrea	Piazza Teatro, 1
.395		Teatro Gian Andrea Dragoni		1838	1978			EmiliaRomagna	Forli-Cesena	Meldola	piazza Felice Orsini, 1
.396		Teatro Gioiello	Pier Giuseppe Mazzarelli	1913				Piemonte	Torino	Torino	Via Cristoforo Colombo, 31
.397		Teatro Giotto		1872	1995			Toscana	Firenze	Borgo San Lorenzo	Corso Giacomo Matteotti, 151
.398		Teatro Giotto	Giuseppe Chiesi	1900	1994			Toscana	Firenze	Vicchio	Piazzetta dei Buoni, 1
.399		Teatro Giovanni Mestica		1903				Marche	Macerata	Apiro	Corso Vittorio Emanuele, 44
.400		Teatro Girolamo Magnani	Nicola Bettoli	1861	1870	2004		EmiliaRomagna	Parma	Fidenza	Piazza Giuseppe Verdi, 1

List of Historical Theatres in Italy

Site of arch. int.	UNESCO Site	ERHT Site	Integrated/Isolated	Construction	Seats	Level	Keyword	Notes	Code
Yes	No	No	Isolated	Wooden and brick	721	Regional			.391
Yes	No	Yes	Isolated	Mixed structure	209	Local			.392
Yes	No	No	Isolated	Mixed structure	220	Local			.393
Yes	No	No	Isolated	Mixed structure	600	Regional			.394
Yes	No	No	Isolated	Mixed structure	318	Local			.395
Yes	No	No	Isolated	Mixed structure	500	Regional		Rebuilt in 1962	.396
Yes	No	No	Isolated	Mixed structure	300	Regional			.397
No	No	No	Isolated	Mixed structure	200	Local			.398
No	No	No	Isolated	Mixed structure	150	Local			.399
Yes	No	No	Isolated	Mixed structure	425	Regional			.400

Code	Code2	Name	Architect/ Engineer	Year	R1	R2	R3	Region	Province	City	Address
.401		Teatro Giuseppe Garibaldi		1700	1998			EmiliaRomagna	Forlì-Cesena	Bagno di Romagna	Via del Teatro, 1
.402		Teatro Giuseppe Piermarini	Giuseppe Piermarini	1812	1852	1873	1995	Marche	Macerata	Matelica	Via Umberto I
.403		Teatro Giuseppe Verdi	Nespega, Siniscalco	1964				Puglia	Brindisi	Brindisi	Via Santi, 1
.404		Teatro Giuseppe Verdi		1823	1989			EmiliaRomagna	Piacenza	Castel San Giovanni	Piazza C. Agostino Casaroli
.405		Teatro Giuseppe Verdi		1873				Marche	Macerata	Pollenza	Piazza della Libertà, 25
.406		Teatro Giuseppe Verdi	Luigi Poletti	1846	1908	1930	1949	Umbria	Terni	Terni	Corso Vecchio, 99
.407		Teatro Gobetti		1842				Piemonte	Torino	Torino	Via Gioachino Rossini, 12
.408		Teatro Goldoni	Giuseppe Del Rosso	1817	1998			Toscana	Firenze	Firenze	Via Santa Maria, 15
.409		Teatro Golfarelli	Giuseppe Missirini	1824	2016			EmiliaRomagna	Forlì-Cesena	Civitella di Romagna	Via G. Garibaldi, 34
.410		Teatro Gonzaga		1800				Lombardia	Cremona	Ostiano	Piazza Garibaldi

Site of arch. int.	UNESCO Site	ERHT Site	Integrated/Isolated	Construction	Seats	Level	Keyword	Notes	Code
Yes	No	No	Isolated	Mixed structure	425	Regional			.400
Yes	No	No	Isolated	Mixed structure	136	Local			.401
Yes	No	No	Isolated	Mixed structure	411	Regional			.402
Yes	No	No	Isolated	Mixed structure	1172	National			.403
Yes	No	No	Isolated	Mixed structure	234	Local			.404
Yes	No	No	Isolated	Wooden and brick	254	Local			.405
Yes	No	No	Isolated	Mixed structure	1066	National			.406
Yes	No	No	Isolated	Wooden and brick		Regional			.407
Yes	Yes	No	Isolated	Mixed structure	363	Regional			.408
Yes	No	No	Isolated	Mixed structure	600	Local			.409
Yes	No	No	Integrated	Wooden and brick		Local		Inside Palazzo Gonzaga	.410

Code	Code2	Name	Architect/Engineer	Year	R1	R2	R3	Region	Province	City	Address
.411		Teatro Gonzaghesco		1852	1919			EmiliaRomagna	Reggio Emilia	Luzzara	Piazza Tedeschi
.412		Teatro Grande	Luigi Canonica	1804				Lombardia	Brescia	Brescia	Via Pagonara, 19
.413		Teatro Grifeo		1862	1993			Sicilia	Palermo	Petralia Sottana	Corso Paola Agliata, 108
.414		Teatro Guglielmi	Vincenzo Micheli	1886	2005			Toscana	Massa-Carrara	Massa	Piazza del Teatro
.415		Teatro Guido	Francesco Piazzalunga	1894	1929	1938	1956	Lombardia	Mantova	Suzzara	Viale S. Zonta, 22
.416		Teatro Gustavo Modena	Nicolò Bruno	1856	1920	1997		Liguria	Genova	Genova	Piazza Gustavo Modena, 3
.417		Teatro Impavidi	Bargigli, Valenti	1809	1816	2005		Liguria	La Spezia	Sarzana	Piazza Garibaldi
.418		Teatro Italia	Angiolo Mazzoni	1930				Lazio	Roma	Roma	Via Bari, 18
.419		Teatro Jacquard		1869				Veneto	Vicenza	Schio	Via Pasubio
.420		Teatro Kursaal		1900				Marche	Macerata	Porto Recanati	Via Fratelli Brancondi

List of Historical Theatres in Italy

Site of arch. int.	UNESCO Site	ERHT Site	Integrated/Isolated	Construction	Seats	Level	Keyword	Notes	Code
Yes	No	No	Isolated	Mixed structure		Local		Damaged by the 2012 earthquake	.411
Yes	No	No	Isolated	Wooden and brick	970	Regional		Altered several times, previous theatre of 1664	.412
Yes	No	No	Isolated	Mixed structure	182	Local			.413
Yes	No	No	Isolated	Mixed structure	448	Regional			.414
Yes	No	No	Isolated	Mixed structure		Local			.415
Yes	No	No	Isolated	Mixed structure	498	Regional			.416
Yes	No	No	Isolated	Mixed structure	350	Regional			.417
Yes	No	No	Integrated	Reinforced concrete	798	Regional			.418
Yes	No	No	Isolated	Wooden and brick	800	Regional		Inside Palazzo di Parco Jacquard	.419
Yes	No	No	Isolated	Wooden and brick	324	Regional			.420

Code	Code2	Name	Architect/Engineer	Year	R1	R2	R3	Region	Province	City	Address
.421		Teatro La Fenice		1734	1991			Marche	Fermo	Amandola	Via Roma
.422		Teatro La Fenice		1910				Lazio	Roma	Arsoli	Piazza Martiri Antifascisti, 1
.423		Teatro La Fenice		1996				Marche	Ancona	Senigallia	Via Cesare Battisti, 19
.424		*Teatro La Fenice*	*Selva, Rossi*	*1789*	*1837*	*2003*		*Veneto*	*Venezia*	*Venezia*	*Campo San Fantin, 1965*
.425		Teatro La Nuova Fenice	Gaetano Canedi	1787	1999			Marche	Ancona	Osimo	Piazza Guglielmo Marconi, 1
.426		Teatro La Rondinella	Daretti, Tombolini	1887	2010			Marche	Macerata	Montefano	Piazza Bracaccini, 1
.427		Teatro La Vittoria	Francesco Fellini	1867	1998			Marche	Ancona	Ostra	Piazza dei Martiri
.428		Teatro Lauro Rossi	Cosimo Morelli	1774				Marche	Macerata	Macerata	Piazza Della Libertà, 21
.429		Teatro Lembo	Arturo Boccasini	1926	2005			Puglia	Barletta-Andria-Trani	Canosa di Puglia	Via Piave, 9
.430		Teatro L'Idea		1851				Sicilia	Agrigento	Sambuca di Sicilia	Corso Umberto I, 34

List of Historical Theatres in Italy

Site of arch. int.	UNESCO Site	ERHT Site	Integrated/Isolated	Construction	Seats	Level	Keyword	Notes	Code
Yes	No	No	Isolated	Mixed structure	287	Local			.421
No	No	No	Isolated	Mixed structure		Local			.422
Yes	No	No	Isolated	Reinforced concrete	874	Regional		Built in place of the previous theatre designed by Ghinelli destroyed by fire	.423
Yes	*Yes*	*Yes*	*Integrated*	*Reinforced concrete*	*1000*	*International*	*Urban node*		*.424*
Yes	No	No	Isolated	Mixed structure		Local			.425
Yes	No	No	Integrated	Mixed structure	150	Local			.426
Yes	No	No	Isolated	Mixed structure	184	Local		Renovation of the previous theatre of 1773	.427
Yes	No	No	Isolated	Wooden and brick	550	Regional			.428
Yes	No	No	Isolated	Mixed structure	408	Regional		Inside Palazzo Comunale	.429
Yes	No	No	Isolated	Wooden and brick	250	Local			.430

Code	Code2	Name	Architect/Engineer	Year	R1	R2	R3	Region	Province	City	Address
.431		Teatro Lirico Giorgio Gaber	Antonio Cassi Ramelli	1939				Lombardia	Milano	Milano	Via Larga, 14
.432		Teatro Lirico Giuseppe Verdi	Pertsch, Selva	1801	1881	1889	1997	FriuliVeneziaGiulia	Trieste	Trieste	Riva 3 Novembre, 1
.433		Teatro Luca Ronconi	Ercole Salmi	1840				Umbria	Perugia	Gubbio	Via Capitano del Popolo, 17
.434		Teatro Luigi Bon		1938				FriuliVeneziaGiulia	Udine	Colugna di Tovagnocco	Via Patrioti, 29
.435		Teatro Luigi Mercantini	Pietro Maggi	1824	1894	2000		Marche	Ascoli Piceno	Ripatransone	Piazza XX Settembre, 11
.436		Teatro Luigi Pirandello	Basile, Sciascia	1880				Sicilia	Agrigento	Agrigento	Piazza Luigi Pirandello, 1
.437		Teatro Malibran	Thomas Bezzi	1678	1834	1992		Veneto	Venezia	Venezia	Cannaregio, 5873
.438		Teatro Manzoni		1895	2002			Toscana	Firenze	Calenzano	Via Mascagni, 18
.439		Teatro Manzoni	Alziro Bergonzo	1850				Lombardia	Milano	Milano	Via Alessandro Manzoni, 42
.440		Teatro Manzoni	Pietro Bernardini	1694				Toscana	Pistoia	Pistoia	Corso Antonio Gramsci, 121

List of Historical Theatres in Italy

Site of arch. int.	UNESCO Site	ERHT Site	Integrated/Isolated	Construction	Seats	Level	Keyword	Notes	Code
Yes	No	Yes	Isolated	Reinforced concrete	1600	National		Reconstruction of the previous Piermarini theatre, La Cannobiana, destroyed in 1938 by a fire	.431
Yes	No	No	Isolated	Mixed structure	1300	National			.432
Yes	No	No	Integrated	Mixed structure		Local		Original building of 1738 by Maurizio Lottici transformed in 1840	.433
No	No	No	Isolated	Reinforced concrete		Regional			.434
Yes	No	No	Isolated	Mixed structure		Local			.435
Yes	No	No	Isolated	Wooden and brick	582	Regional			.436
Yes	Yes	No	Isolated	Mixed structure		Regional			.437
Yes	No	No	Isolated	Mixed structure	142	Local			.438
Yes	No	Yes	Isolated	Reinforced concrete	850	Regional		Reconstruction of the previous theatre designed by Andrea Scala in 1870, destroyed by bombing in 1943	.439
Yes	No	No	Isolated	Mixed structure	836	Regional			.440

Code	Code2	Name	Architect/Engineer	Year	R1	R2	R3	Region	Province	City	Address
441		Teatro Maratti		1870	1913	1933		Marche	Ancona	Camerano	Piazza Roma, 31
442		Teatro Marchetti	Vincenzo Ghinelli	1847	1905	1990		Marche	Macerata	Camerino	Corso Vittorio Emanuele II, 17
443		Teatro Marchinneschi		1833	1982			Toscana	Pisa	Guardistallo	Via Palestro, 44
444		Teatro Marenco	Donato Levi	1860				Marche	Macerata	Muccia	Via Roma, 32
445		Teatro Margherita	Francesco De Giuglio	1914				Puglia	Bari	Bari	Piazza Quattro Novembre
446		Teatro Marinoni		1923				Veneto	Venezia	Lido di Venezia	Lungomare G. D' Annunzio
447		Teatro Mario Tiberini	Luigi Tiberini	1816	1880	1981		Marche	Pesaro-Urbino	San Lorenzo in Campo	Via Mario Tiberini, 3
448		Teatro Marrucino	Eugenio Michitelli	1818	1872			Abruzzo	Chieti	Chieti	Via Cesare de Lollis, 1
449		Teatro Martinetti		1833	1977			Lombardia	Pavia	Garlasco	Via Santissima Trinità, 6
450		Teatro Mascagni		1838	1996			Toscana	Siena	Chiusi	Via Garibaldi

List of Historical Theatres in Italy

Site of arch. int.	UNESCO Site	ERHT Site	Integrated/Isolated	Construction	Seats	Level	Keyword	Notes	Code
Yes	No	No	Isolated	Mixed structure		Local			.441
Yes	No	No	Isolated	Mixed structure	520	Regional			.442
Yes	No	No	Isolated	Mixed structure	183	Local			.443
Yes	No	No	Isolated	Wooden and brick		Local			.444
Yes	No	No	Isolated	Reinforced concrete		Regional		Conversion into a museum	.445
Yes	No	No	Isolated	Mixed structure		Regional			.446
Yes	No	No	Isolated	Mixed structure	150	Local			.447
Yes	No	No	Isolated	Mixed structure	460	Regional			.448
Yes	No	No	Isolated	Mixed structure	255	Local			.449
Yes	No	No	Isolated	Mixed structure	400	Regional			.450

Code	Code2	Name	Architect/ Engineer	Year	R1	R2	R3	Region	Province	City	Address
.451		Teatro Mascagni		1930	in corso			Campania	Caserta	Piedimonte Matese	Via Antonio de Cesare
.452		Teatro Mascagni		1890	1984	1998		Toscana	Pistoia	San Marcello Piteglio	Via delle Corti
.453		Teatro Massimo Bontempelli		19th cent	1917			Umbria	Perugia	Citerna	Via della Rocca
.454		Teatro Massimo Vittorio Emanuele	E. Basile, G. B. F. Basile	1897	1997			Sicilia	Palermo	Palermo	Piazza Verdi
.455		Teatro Mattarello		1904	1994			Veneto	Vicenza	Arzignano	Corso Giuseppe Mazzini, 22
.456		Teatro Mazzini		1911				EmiliaRomagna	Ravenna	Castiglione di Cervia	Via Zattoni, 34
.457		Teatro Mediceo	Bernardo Buontalenti	1856				Toscana	Firenze	Firenze	
.458		Teatro Mengoni	Giuseppe Mengoni	1870				Umbria	Perugia	Magione	Piazza Mengoni, 1A
.459		Teatro Mercadante	Vincenzo Striccoli	1895	2003			Puglia	Bari	Altamura	Via Selva, 101
.460		Teatro Mercadante	Leopoldo Vaccaro	1868	1900	1939	1994	Puglia	Foggia	Cerignola	Piazza Giacomo Matteotti, 1

List of Historical Theatres in Italy

Site of arch. int.	UNESCO Site	ERHT Site	Integrated/Isolated	Construction	Seats	Level	Keyword	Notes	Code
No	No	No	Isolated	Reinforced concrete		Local			.451
Yes	No	No	Isolated	Mixed structure	130	Local			.452
No	No	No	Isolated	Mixed structure	90	Local			.453
Yes	No	No	Isolated	Mixed structure	1381	International			.454
No	No	No	Integrated	Mixed structure	432	Regional		Inside a Venetian villa	.455
Yes	No	No	Isolated	Mixed structure		Regional			.456
Yes	Yes	No	Integrated	Wooden and brick		National		Inside Palazzo degli Uffizi, closed in the eighteenth century	.457
Yes	No	No	Isolated	Wooden and brick	238	Local			.458
Yes	No	No	Isolated	Mixed structure	500	Regional			.459
Yes	No	No	Isolated	Mixed structure	300	Regional			.460

Code	Code2	Name	Architect/Engineer	Year	R1	R2	R3	Region	Province	City	Address
.461		Teatro Mercadante	Francesco Securo	1777	1893			Campania	Napoli	Napoli	Piazza Francese, 46
.462		Teatro Metastasio	Luigi de Cambray Digny	1830				Toscana	Prato	Prato	Via Benedetto Cairoli, 59
.463		Teatro Michetti		1948				Abruzzo	Pescara	Pescara	Via D'Annunzio, 28
.464		Teatro Misa	Ferroni, Ghinelli	1840				Marche	Ancona	Arcevia	Corso Mazzini 22
.465		Teatro Modernissimo	Gualtiero Pontoni	1914	2016			EmiliaRomagna	Bologna	Bologna	Via Francesco Rizzoli, 1
.466		Teatro Modernissimo		1910	1985			Veneto	Vicenza	Noventa Vicentina	Via Broli
.467		Teatro Moderno		1910				Campania	Napoli	Torre Annunziata	Piazza Nicotera
.468		Teatro Morelli		1930	2008			Calabria	Cosenza	Cosenza	Via Lungo B. Oberdan, 37
.469		Teatro Morlacchi	Alessio Lorenzini	1781	1874	1956		Umbria	Perugia	Perugia	Piazza Morlacchi, 13
.470		Teatro Mugellini	Giuseppe Brandoni	1862				Marche	Macerata	Potenza Picena	Piazza Giacomo Matteotti, 1

List of Historical Theatres in Italy

Site of arch. int.	UNESCO Site	ERHT Site	Integrated/Isolated	Construction	Seats	Level	Keyword	Notes	Code
Yes	Yes	No	Isolated	Mixed structure	553	Regional			.461
Yes	No	No	Isolated	Wooden and brick	686	Regional			.462
Yes	No	No	Isolated	Reinforced concrete	850	Regional		Reconstruction of the 1910 theatre destroyed by a fire in 1944	.463
Yes	No	No	Integrated	Wooden and brick	157	Local		Inside Palazzo dei Priori	.464
Yes	No	No	Isolated	Reinforced concrete	400	Regional			.465
No	No	No	Isolated	Reinforced concrete	296	Local		Reconstruction of the theatre of 1876	.466
No	No	No	Isolated	Wooden and brick		Regional		Closed due to unavailability in the 80s	.467
Yes	No	No	Isolated	Reinforced concrete	626	Regional			.468
Yes	No	No	Isolated	Mixed structure	785	Regional			.469
Yes	No	No	Isolated	Wooden and brick	99	Local			.470

Code	Code2	Name	Architect/ Engineer	Year	R1	R2	R3	Region	Province	City	Address
.471		Teatro Municipale	Agostino Vitoli	1791	1840	1990		Piemonte	Alessandria	Casale Monferrato	Piazza Castello, 9
.472		Teatro Municipale	Marco Zanuso	1999				TrentinoAltoAdige	Bolzano	Bolzano	Piazza Giuseppe Verdi, 40
.473		Teatro Municipale	Monti, Santini	1876				Marche	Pesaro–Urbino	Cagli	Piazza Papa Niccolò IV
.474		Teatro Municipale	Filippo Amici	1901	1980			Marche	Macerata	Caldarola	Piazza Mattei, 1
.475		Teatro Municipale		1874	2006			Puglia	Bari	Corato	Piazza Marconi, 7
.476		Teatro Municipale	Luigi Catalani	1857				Abruzzo	L'aquila	L'aquila	Piazza del Teatro
.477		Teatro Municipale	Lotario Tomba	1804	1857	1938	1976	EmiliaRomagna	Piacenza	Piacenza	Via Verdi, 41
.478		Teatro Municipale	Tommaso Onofrio di Canelli	1836	1991			Piemonte	Torino	Pinerolo	Piazza Vittorio Veneto, 1
.479		Teatro Municipale	Giuseppe Locatelli	1817				Marche	Fermo	Porto San Giorgio	Largo del Teatro, 1
.480		Teatro Municipale	Cesare Costa	1857				EmiliaRomagna	Reggio Emilia	Reggio Emilia	Piazza Martiri 7 Luglio

List of Historical Theatres in Italy

Site of arch. int.	UNESCO Site	ERHT Site	Integrated/Isolated	Construction	Seats	Level	Keyword	Notes	Code
Yes	No	No	Isolated	Mixed structure	500	Regional			.471
Yes	No	No	Isolated	Reinforced concrete	814	Regional			.472
Yes	No	No	Isolated	Wooden and brick	501	Regional			.473
Yes	No	No	Isolated	Mixed structure	291	Local			.474
Yes	No	No	Isolated	Mixed structure		Local			.475
Yes	No	No	Isolated	Mixed structure	600	Regional			.476
Yes	No	No	Isolated	Mixed structure	1174	National			.477
Yes	No	No	Isolated	Mixed structure	560	Regional		Rebuilt, original facade	.478
Yes	No	No	Isolated	Wooden and brick	278	Local			.479
Yes	No	No	Isolated	Wooden and brick	1136	National			.480

Code	Code2	Name	Architect/Engineer	Year	R1	R2	R3	Region	Province	City	Address
.481		Teatro Municipale Ballarin		1814	1915	1948	2002	Veneto	Rovigo	Lendinara	Via Gianbattista Conti, 4
.482		Teatro Municipale Giuseppe Verdi	D'Amora, Menechini	1872				Campania	Salerno	Salerno	Piazza Matteo Luciani
.483		Teatro Naselli	Battaglia, Sortino	1826	1970	2006		Sicilia	Ragusa	Comiso	Via Giuseppe Morso
.484		Teatro Nazionale		1789	1840			Toscana	Firenze	Firenze	Via dei Cimatori, 6
.485		Teatro Nazionale	Mauro Rota	1924	1979	2001	2009	Lombardia	Milano	Milano	Piazza Piemonte, 12
.486		Teatro Niccolini		1648	1764	1914	2007	Toscana	Firenze	Firenze	Via Ricasoli, 3/5
.487		Teatro Niccolini	Antonio Sodi	1850	1938	1996		Toscana	Firenze	San Casciano in Val di Pesa	Piazza della Repubblica, 12
.488		Teatro Nicola Antonio Angeletti		1883	1998			Marche	Macerata	Sant' Angelo in Pontano	Via San Nicola
.489		Teatro Nicola degli Angeli	Giuseppe Sabbatii	1898				Marche	Macerata	Montelupone	Piazza del Comune, 1
.490		Teatro Nicola Vaccaj	Giuseppe Lucatelli	1797	1882	1985	2008	Marche	Macerata	Tolentino	Piazza Vaccaj

List of Historical Theatres in Italy

Site of arch. int.	UNESCO Site	ERHT Site	Integrated/Isolated	Construction	Seats	Level	Keyword	Notes	Code
Yes	No	No	Isolated	Mixed structure	453	Regional			.481
Yes	No	No	Isolated	Wooden and brick	610	Regional			.482
Yes	No	No	Isolated	Mixed structure	264	Local			.483
Yes	Yes	No	Isolated	Mixed structure		Regional			.484
Yes	No	Si	Isolated	Reinforced concrete	1500	national			.485
Yes	Yes	No	Isolated	Mixed structure	500	Regional			.486
Yes	No	No	Isolated	Mixed structure	344	Regional			.487
Yes	No	No	Isolated	Mixed structure	100	Local			.488
Yes	No	No	Isolated	Wooden and brick	272	Local			.489
Yes	No	No	Isolated	Reinforced concrete	478	Regional			.490

Code	Code2	Name	Architect/Engineer	Year	R1	R2	R3	Region	Province	City	Address
491		Teatro Novissimo	Giacomo Torelli	1640				Veneto	Venezia	Venezia	
492		Teatro Nuovo	Sesto Boari	1926	2015			EmiliaRomagna	Ferrara	Ferrara	Piazza Trento Trieste, 52
493		Teatro Nuovo		1779				Toscana	Firenze	Firenze	Via Maurizio Bufalini, 9
494		Teatro Nuovo	Lorenzo Colliva	1904	in corso			EmiliaRomagna	Modena	Mirandola	Piazza Della Costituente, 7
495		Teatro Nuovo		1985				Campania	Napoli	Napoli	Via Montecalvario, 16
496		Teatro Nuovo	Enrico Storari	1846				Veneto	Verona	Verona	Piazza Viviani, 10
497		Teatro Nuovo Andreani		1862				Lombardia	Mantova	Mantova	Corso Vittorio Emanuele II, 73
498		Teatro Nuovo Gian Carlo Menotti	Ireneo Aleandri	1864	2007			Umbria	Perugia	Spoleto	Via Vaita S. Andrea
499		Teatro Nuovo Mario Monicelli	Antonio Foglia	1830	1912			Lombardia	Mantova	Ostiglia	Via Ghinosi, 18
500	CODE 3.01 and 3.24	Teatro Olimpico	Andrea Palladio	1585	1986			Veneto	Vicenza	Vicenza	Piazza Matteotti, 1

Site of arch. int.	UNESCO Site	ERHT Site	Integrated/Isolated	Construction	Seats	Level	Keyword	Notes	Code
Yes	Yes	No	Isolated	wooden		Regional		Demolished in 1645	.491
Yes	Yes	No	Isolated	Reinforced concrete	800	Regional			.492
Yes	Yes	No	Isolated	Wooden and brick		Regional		Change of intended use, renovation in 2003 based on a project by Natalini	.493
Yes	No	No	Isolated	Reinforced concrete	409	Regional		Unavailable	.494
Yes	Yes	No	Isolated	Reinforced concrete		Regional		Original theatre since 1724 destroyed by fire in 1861, rebuilt and destroyed by fire in 1935	.495
Yes	Yes	No	Isolated	Wooden and brick	884	National			.496
Yes	Yes	Yes	Isolated	Wooden and brick		Regional		Change of intended use	.497
Yes	No	No	Isolated	Mixed structure	800	Regional	Urban node		.498
Yes	No	No	Isolated	Mixed structure		Local			.499
Yes	Yes	Yes	Integrated	Wooden and brick	470	International	Organism		.500

Code	Code2	Name	Architect/Engineer	Year	R1	R2	R3	Region	Province	City	Address
.501		Teatro Orfeo		1815				Puglia	Taranto	Taranto	Via Pitagora, 78
.502		Teatro Pacini	Giovanni Antonio	1728	1795	1842	1949	Toscana	Pistoia	Pescia	Piazza San Francesco
.503		Teatro Paisello		1872	1999			Puglia	Lecce	Lecce	Via Giuseppe Palmieri
.504		Teatro Palazzo Litta		1750				Lombardia	Milano	Milano	Corso Magenta, 24
.505		Teatro Pallavicino		1804	1827	1910	2003	EmiliaRomagna	Parma	Zibello	Piazza Garibaldi
.506		Teatro Paolella	Benedetto Novellis	1760	2009			Calabria	Cosenza	Rossano	Via Teatro
.507		Teatro Paolo Ferrari	Enrico Medi	1871	1930			Marche	Ancona	San Marcello	Via Rossetti
.508		Teatro Parrocchiale		1900				Lombardia	Brescia	Cellatica	Via Montebello, 12
.509		Teatro Pergolesi	Cosimo Morelli	1790				Marche	Ancona	Jesi	Piazza della Repubblica, 9
.510		Teatro Persiani	Tommaso Bandoni	1840	2003			Marche	Macerata	Recanati	Via C. B. Conte di Cavour

List of Historical Theatres in Italy

Site of arch. int.	UNESCO Site	ERHT Site	Integrated/Isolated	Construction	Seats	Level	Keyword	Notes	Code
Yes	No	No	Isolated	Reinforced concrete	527	Regional			.501
Yes	No	No	Isolated	Mixed structure	445	Regional			.502
Yes	No	No	Isolated	Mixed structure	320	Regional			.503
Yes	No	No	Isolated	Wooden and brick	200	Local			.504
Yes	No	No	Integrated	Mixed structure		Local		Inside Palazzo Litta	.505
Yes	No	No	Isolated	Mixed structure		Local			.506
Yes	No	No	Integrated	Wooden and brick	143	Local			.507
No	No	No	Isolated	Wooden and brick		Local			.508
Yes	No	No	Isolated	Wooden and brick	712	Regional			.509
Yes	No	No	Isolated	Mixed structure	400	Regional			.510

Code	Code2	Name	Architect/Engineer	Year	R1	R2	R3	Region	Province	City	Address
511		Teatro Persio Flacco	Luigi Campani	1816	1999			Toscana	Pisa	Toscana	Via dei Sarti, 37
512		Teatro Perugini		1876	2000			Marche	Pesaro-Urbino	Apecchio	Piazza San Martino
513		Teatro Petrarca	Vittorio Bellini	1833	1882	2010		Toscana	Arezzo	Arezzo	Via Guido Monaco, 12
514		Teatro Petruzzelli	Angelo Cicciomessere	1905	2009			Puglia	Bari	Bari	Corso Cavour, 12
515		Teatro Piccinni	Antonio Niccolini	1854				Puglia	Bari	Bari	Corso Vittorio Emanuele II, 84
516		Teatro Piccolo Eliseo		1900				Lazio	Roma	Roma	Via Nazionale, 183
517		Teatro Pisciotta		1900				Veneto	Verona	Casaleone	Via Vittorio Veneto, 61
518		Teatro Politeama	Saverio Fragapane	1900				Sicilia	Catania	Caltagirone	Via Giardini Pubblici, 4
519		Teatro Politeama	Federico Frigerio	1910				Lombardia	Como	Como	Via Tolomeo Gallio, 38
520		Teatro Politeama	Oronzo Greco	1884	1913	1926		Puglia	Lecce	Lecce	Via XXV Luglio, 30

List of Historical Theatres in Italy

Site of arch. int.	UNESCO Site	ERHT Site	Integrated/Isolated	Construction	Seats	Level	Keyword	Notes	Code
Yes	No	No	Isolated	Mixed structure	500	Regional			.511
Yes	No	No	Integrated	Mixed structure	42	Local		Inside Palazzo Ubaldini	.512
Yes	No	No	Isolated	Mixed structure	698	Regional			.513
Yes	No	No	Isolated	Reinforced concrete	1250	International		Damaged by fire in 1991	.514
Yes	No	No	Isolated	Wooden and brick	620	Regional			.515
Yes	Yes	No	Integrated	Mixed structure	265	Local			.516
No	No	No	Isolated	Reinforced concrete		Local		Closed due to unavailability in the 60s	.517
Yes	No	No	Isolated	Wooden and brick		Local			.518
Yes	No	No	Isolated	Reinforced concrete	1300	National			.519
Yes	No	No	Isolated	Mixed structure	998	National			.520

Code	Code2	Name	Architect/Engineer	Year	R1	R2	R3	Region	Province	City	Address
.521		Teatro Politeama		1961				Campania	Napoli	Napoli	Via Monte di Dio, 80
.522		Teatro Politeama		1911				Lombardia	Mantova	Suzzara	Strada Zara Bignardina, 4
.523		Teatro Politeama		1949	2000			Toscana	Lucca	Viareggio	Largo Molo Del Greco Corrado
.524		Teatro Politeama Boglione	Achille Sfondrini	1900				Piemonte	Cuneo	Bra	Piazza Carlo Alberto, 23
.525		Teatro Politeama Garibaldi	Giuseppe Damiani Almeyda	1875	1896	2000		Sicilia	Palermo	Palermo	Piazza Ruggero Settimo, 15
.526		Teatro Politeama Giuseppe Verdi	Leandro Caselli	1892	1983	2003		Toscana	Massa-Carrara	Carrara	Piazza Giacomo Matteotti
.527		Teatro Politeama Rossetti	Nicolò Bruno	1878	1928	1969	2001	FriuliVeneziaGiulia	Trieste	Trieste	Viale XX Settembre, 45

(continued)

List of Historical Theatres in Italy

(continued)

Code	Code2	Name	Architect/Engineer	Year	R1	R2	R3	Region	Province	City	Address
.528		Teatro Politeama Verdi	Achille Sfondrini	1898				Lombardia	Cremona	Cremona	Via Cesare Battisti, 3
.529		Teatro Poliziano	Giuseppe Valentini	1795	1870	1945	1980	Toscana	Siena	Montepulciano	Via del Teatro, 4
.530		Teatro Popolare d'Arte		1900				Toscana	Arezzo	Bucine	Via del Teatro, 14

Site of arch. int.	UNESCO Site	ERHT Site	Integrated/Isolated	Construction	Seats	Level	Keyword	Notes	Code
Yes	Yes	No	Isolated	Reinforced concrete	900	Regional		Reconstruction of the previous theatre of 1872 destroyed by a fire in 1957	.521
Yes	No	No	Isolated	Reinforced concrete	100	Local			.522
Yes	No	No	Isolated	Reinforced concrete	1840	National		Reconstruction of the previous theatre of 1869 destroyed by the bombing of 1943	.523
No	No	No	Isolated	Wooden and brick	650	Regional			.524
Yes	No	No	Isolated	Mixed structure	950	Local			.525
Yes	No	No	Isolated	Mixed structure	1500	National		Closed due to unavailability	.526
Yes	No	No	Isolated	Mixed structure	1531	National			.527
Yes	No	No	Isolated	Wooden and brick		Regional			.528
Yes	No	No	Isolated	Mixed structure	426	Regional			.529
No	No	No	Isolated	Wooden and brick	265	Local			.530

Code	Code2	Name	Architect/Engineer	Year	R1	R2	R3	Region	Province	City	Address
531		Teatro Puccini	Martin Dülfer	1900				TrentinoAltoAdige	Bolzano	Merano	Piazza Teatro, 2
532		Teatro Quirino	De Angelis, Marra	1871	1882	1914	1954	Lazio	Roma	Roma	Via delle Vergini, 7
533		Teatro Rasi		1892	1962			EmiliaRomagna	Ravenna	Ravenna	Via di Roma, 39
534		*Teatro Re*	*Luigi Canonica*	*1813*				*Lombardia*	*Milano*	*Milano*	*Via Silvio Pellico*
535		Teatro Re Grillo	Filippo Re Grillo	1923	2003			Sicilia	Agrigento	Licata	Corso Vittorio Emanuele, 53
536		Teatro Regina Margherita		1750	2003			Toscana	Siena	Barberino Val d'Elsa	Via A. Mori, 28
537		Teatro Regina Margherita		1875	1921	1973		Sicilia	Caltanissetta	Caltanissetta	Corso Vittorio Emanuele II, 1
538		Teatro Regina Margherita	Dionisio Sciascia	1880	2003			Sicilia	Agrigento	Racalmuto	Via Vittorio Emanuele, 10–6
539		Teatro Regio	Nicola Bettoli	1821	1853			EmiliaRomagna	Parma	Parma	Via Garibaldi, 16A
540		*Teatro Regio*	*Alfieri, Cocito, Mollino*	*1761*	*1881*	*1973*		*Piemonte*	*Torino*	*Torino*	*Piazza Castello, 215*

Site of arch. int.	UNESCO Site	ERHT Site	Integrated/Isolated	Construction	Seats	Level	Keyword	Notes	Code
Yes	No	No	Isolated	Wooden and brick	296	Local			.531
Yes	Yes	No	Isolated	Mixed structure	852	Regional			.532
Yes	No	No	Isolated	Mixed structure	892	Regional			.533
Yes	*No*	*Yes*	*Isolated*	*Wooden and brick*		*Regional*		*Demolished in 1872*	*.534*
Yes	No	No	Integrated	Reinforced concrete		Local			.535
Yes	No	No	Isolated	Mixed structure	90	Local			.536
Yes	No	No	Isolated	Mixed structure	450	Regional			.537
Yes	No	No	Isolated	Mixed structure	350	Regional			.538
Yes	No	Yes	Isolated	Mixed structure	1092	International		Previous theatre of 1697 demolished at the beginning of the nineteenth century	.539
Yes	*No*	*No*	*Integrated*	*Reinforced concrete*	*1592*	*International*	*Urban node*	*Original facade, former theatre destroyed by fire in 1936*	*.540*

List of Historical Theatres in Italy

Code	Code2	Name	Architect/Engineer	Year	R1	R2	R3	Region	Province	City	Address
.541	CODE 3.25	Teatro Regio di Mantova	Bertani, Viani, Bibiena, Piermarini	1549	1783			Lombardia	Mantova	Mantova	Piazza Sordello, 40
.542		Teatro Ricciardi	Francesco Gasperi	1781	1929			Campania	Caserta	Capua	Largo Porta Napoli
.543		Teatro Ristori		1837	2011			Veneto	Verona	Verona	Vicolo Valle 2/A, 7
.544		Teatro Roma		1846				Toscana	Livorno	Castagneto Carducci	Via Antonio Gramsci, 3
.545		Teatro Romualdo Marenco	Giuseppe Becchi	1839	2018			Piemonte	Alessandria	Novi Ligure	Via Municipio
.546		Teatro Ronci		1880				EmiliaRomagna	Rimini	Morciano di Romagna	Via Roma, 60
.547		Teatro Rosaspina		1800				EmiliaRomagna	Rimini	Montescudo	Piazza del Municipio, 1
.548		Teatro Rosini		1790	1861	1954	1987	Toscana	Arezzo	Lucignano	Via Rosini
.549		Teatro Rossetti	Taddeo Salvini	1819	1841	1905	1987	Abruzzo	Chieti	Vasto	Via Aimone, 1
.550		Teatro Rossini	Felice Ravillon	1843	1899	1959	1987	Puglia	Bari	Gioia del Colle	Via Gioacchino Rossini, 1

Site of arch. int.	UNESCO Site	ERHT Site	Integrated/Isolated	Construction	Seats	Level	Keyword	Notes	Code
Yes	*Yes*	*Yes*	*Isolated*	*Wooden and brick*		*National*	*Organism*	*Demolished, archaeological museum in its place*	*.541*
Yes	No	No	Isolated	Mixed structure		Local		Formerly the sixteenth century Teatro Pubblico Campano	.542
Yes	Yes	No	Isolated	Mixed structure	496	Regional			.543
Yes	No	No	Isolated	Wooden and brick	190	Local			.544
Yes	No	No	Isolated	Mixed structure		Local			.545
Yes	No	No	Isolated	Wooden and brick	400	Regional			.546
Yes	No	No	Isolated	Wooden and brick		Local			.547
Yes	No	No	Isolated	Mixed structure		Local			.548
Yes	No	No	Isolated	Mixed structure	156	Local			.549
Yes	No	No	Isolated	Mixed structure	600	Regional			.550

List of Historical Theatres in Italy 333

Code	Code2	Name	Architect/ Engineer	Year	R1	R2	R3	Region	Province	City	Address
.551		Teatro Rossini	Bibiena, Petrocchi	1761	1921			EmiliaRomagna	Ravenna	Lugo	Piazzale Cavour, 17
.552		Teatro Rossini		1818	1934	1980	2002	Marche	Pesaro-Urbino	Pesaro	Piazza Lazzarini, 1
.553		Teatro Rossini	Virginio Vespignani	1874	1950			Lazio	Roma	Roma	Piazza di Santa Chiara, 14
.554		Teatro Rossini		1900				Toscana	Pisa	San Giuliano Terme	Piazza Palmiro Togliatti
.555		Teatro Ruggero		1850				Basilicata	Potenza	Melfi	Via Vittorio Emanuele, 29
.556		Teatro Ruggieri		1671	1814	1965		EmiliaRomagna	Reggio Emilia	Guastalla	Piazza Mazzini
.557		Teatro Sacco		1785	1888			Liguria	Savona	Savona	Via Quarda Superiore, 1
.558		Teatro Sala Umberto	Andrea Busiri Vici	1882	1913	1981		Lazio	Roma	Roma	Via della Mercede, 50
.559		Teatro Salieri	Bressan, Maggioni	1912–1956	1986			Veneto	Verona	Legnago	Via XX Settembre, 26
.560		Teatro Salinus	Giuseppe Patricolo	1908				Sicilia	Trapani	Castelvetrano	Piazzale Carlo D'Aragona, 7

Site of arch. int.	UNESCO Site	ERHT Site	Integrated/Isolated	Construction	Seats	Level	Keyword	Notes	Code
Yes	No	No	Isolated	Mixed structure	448	Regional			.551
Yes	No	No	Isolated	Mixed structure	860	Regional		Reconstruction of the previous Teatro del Sole of 1637	.552
Yes	Yes	No	Integrated	Mixed structure	200	Local		Altered at the end of the nineteenth century, partly restored in the 50s	.553
Yes	No	No	Isolated	Wooden and brick	216	Local			.554
Yes	No	No	Isolated	Wooden and brick		Local			.555
Yes	No	No	Isolated	Mixed structure	430	Regional			.556
Yes	No	No	Isolated	Mixed structure	300	Regional			.557
Yes	Yes	No	Isolated	Mixed structure	485	Regional			.558
Yes	No	No	Isolated	Reinforced concrete	630	Regional			.559
Yes	No	No	Isolated	Mixed structure		Local			.560

List of Historical Theatres in Italy

Code	Code2	Name	Architect/ Engineer	Year	R1	R2	R3	Region	Province	City	Address
.561		Teatro Salvini		1830	2004			Liguria	Imperia	Pieve di Teco	Via Umberto I, 4
.562		Teatro Salvini		1823	1870	1934	1972	Toscana	Grosseto	Pitigliano	Piazza Garibaldi, 37
.563		Teatro San Babila	Gottardi, Danova Belisario Duca	1964				Lombardia	Milano	Milano	Piazza San Babila
.564		Teatro San Bartolomeo		1683				Campania	Napoli	Napoli	Via San Bartolomeo, 6
.565		Teatro San Benedetto	Pietro Checchia	1755				Veneto	Venezia	Venezia	Salita del Teatro
.566		Teatro San Carlo	Medrano, Nicolini	1737	1816	2009		Campania	Napoli	Napoli	Via San Carlo, 98F
.567		Teatro San Cassiano		1637				Veneto	Venezia	Venezia	
.568		Teatro San Ferdinando	Camillo Lionti	1791	1971	2007		Campania	Napoli	Napoli	Piazza Eduardo de Filippo, 20
.569		Teatro San Giovanni Bosco		1900				Lombardia	Cremona	Capergnanica	Via Mons. Antonietti, 18
.570		Teatro San Luigi	Vincenzo Pantoli	1893	1965	2000		EmiliaRomagna	Forlì-Cesena	Forlì	Via L. Nanni, 14

Site of arch. int.	UNESCO Site	ERHT Site	Integrated/Isolated	Construction	Seats	Level	Keyword	Notes	Code
Yes	No	No	Isolated	Mixed structure		Local			.561
Yes	No	No	Isolated	Mixed structure	99	Local			.562
Yes	No	No	Integrated	Reinforced concrete	474	Regional	Knotting		.563
Yes	Yes	No	Isolated	Wooden and brick		Regional		Converted into a church in 1740	.564
Yes	Yes	No	Integrated	Wooden and brick		Regional		Destroyed by fire in 1793, rebuilt, altered in 1937 and closed in 2007	.565
Yes	Yes	No	Isolated	Mixed structure	1386	International	Urban node	Destroyed by fire in 1816, rebuilt and damaged by the bombing of 1943, rebuilt	.566
Yes	Yes	Yes	Integrated	Wooden and brick		Regional	Knotting	Closed in 1807 by Napoleonic decree	.567
Yes	Yes	No	Isolated	Mixed structure	500	Regional			.568
No	No	No	Isolated	Wooden and brick		Local			.569
Yes	No	No	Isolated	Mixed structure		Regional			.570

List of Historical Theatres in Italy

Code	Code2	Name	Architect/Engineer	Year	R1	R2	R3	Region	Province	City	Address
.571		Teatro San Marco	Pampaloni, Piccioli	1806	2017			Toscana	Livorno	Livorno	Via dei Floridi, 9
.572		Teatro San Maritno		1874	2013			Sardegna	Oristano	Oristano	Via Ciutadella De Menorca, 21
.573		Teatro San Moise		1640				Veneto	Venezia	Venezia	
.574		Teatro San Salvatore		1520				EmiliaRomagna	Bologna	Bologna	Via Volto Santo,1
.575		*Teatro San Samuele*		*1656*				*Veneto*	*Venezia*	*Venezia*	
.576		Teatro San Teodoro Cantù		1921	2011			Lombardia	Como	Cantù	Via Eugenio Corbetta, 7
.577		Teatro Sandro Palmieri		1882				Liguria	Imperia	Diano Marina	Via Cairoli, 35
.578		Teatro Sangiorgi	Salvatore Giuffrida	1900	2002			Sicilia	Catania	Catania	Via Antonino di Sangiuliano, 233
.579		Teatro Sannazzaro	Fausto Niccolini	1884	1971			Campania	Napoli	Napoli	Via Chiaia, 157
.580		Teatro Sant' Alessandro		1900				Sicilia	Agrigento	Santa Margherita di Belice	Piazza Matteotti

Site of arch. int.	UNESCO Site	ERHT Site	Integrated/Isolated	Construction	Seats	Level	Keyword	Notes	Code
Yes	No	No	Isolated	Mixed structure		Regional			.571
Yes	No	No	Isolated	Mixed structure		Regional			.572
Yes	Yes	No	Isolated	Wooden and brick		Regional		Closed in 1818 and converted into a building with shops	.573
Yes	No	No	Integrated	Wooden and brick	99	Local			.574
Yes	*Yes*	*No*	*Isolated*	*Wooden and brick*		*Regional*		*Destroyed by fire in 1747, rebuilt and demolished in 1889*	.575
Yes	No	No	Isolated	Reinforced concrete	200	Local			.576
Yes	No	No	Isolated	Wooden and brick	268	Local			.577
Yes	No	No	Isolated	Mixed structure	477	Regional			.578
Yes	Yes	No	Isolated	Mixed structure		Regional			.579
No	No	No	Isolated	Wooden and brick	300	Local			.580

List of Historical Theatres in Italy

Code	Code2	Name	Architect/Engineer	Year	R1	R2	R3	Region	Province	City	Address
581		Teatro Santi Giovanni e Paolo	Carlo Fontana	1638				Veneto	Venezia	Venezia	Calle della Testa
582		Teatro Sanzio	Ghinelli, De Carlo	1845	1977			Marche	Pesaro-Urbino	Pesaro	Via Giacomo Matteotti
583		Teatro Savoia		1926	1933			Molise	Campobasso	Campobasso	Piazza Gabriele Pepe, 5
584		Teatro Scientifico Bibiena	Antonio Galli Bibiena	1767				Lombardia	Mantova	Mantova	Via Accademica, 47
585		Teatro Srcoffa	Francesco Mazzarelli	1692				EmiliaRomagna	Ferrara	Ferrara	
586		Teatro Selve	Domenico Berutto	1855	1962	2007		Piemonte	Torino	Vigone	Vicolo del Teatro
587		Teatro Servadio		1850	1990			Toscana	Siena	Abbadia San Salvatore	Via Pinelli, 30
588		Teatro Signorelli	Carlo Gatteschi	1854				Toscana	Arezzo	Cortona	Piazza Luca Signorelli, 3
589		Teatro Sivori	Giorgio Finocchio	1824	in corso			Liguria	Savona	Finale Ligure	Via Torino, 3
590		Teatro Sociale	Stefano Cansacchi	1783	1866	1886	2006	Umbria	Terni	Amelia	Via del Teatro, 39

Site of arch. int.	UNESCO Site	ERHT Site	Integrated/Isolated	Construction	Seats	Level	Keyword	Notes	Code
Yes	*Yes*	*No*	*Isolated*	*Wooden and brick*	*1000*	*national*		*Demolished in 1715 after the roof collapsed*	*.581*
Yes	No	No	Isolated	Mixed structure	460	Regional			.582
Yes	No	No	Integrated	Reinforced concrete		Regional		Built on the previous Teatro Margherita of 1894	.583
Yes	*Yes*	*Yes*	*Integrated*	*Wooden and brick*	*338*	*Regional*	*Knotting*		*.584*
Yes	*Yes*	*No*	*Isolated*	*Wooden and brick*		*Regional*	*Organism*	*Demolished in 1810*	*.585*
Yes	No	No	Isolated	Mixed structure	142	Local			.586
Yes	No	No	Isolated	Mixed structure	95	Local			.587
Yes	No	No	Isolated	Wooden and brick		Local			.588
Yes	No	No	Isolated	Mixed structure	400	Regional			.589
Yes	No	No	Isolated	Mixed structure	380	Local			.590

Code	Code2	Name	Architect/Engineer	Year	R1	R2	R3	Region	Province	City	Address
.591		Teatro Sociale		1891				Lombardia	Mantova	Asola	Via Piave, 17
.592		Teatro Sociale		1879				Piemonte	Vercelli	Balmuccia	Via Baraggiolo
.593		Teatro Sociale	Leopoldo Pollack	1807	1981	2009		Lombardia	Bergamo	Bergamo	Via B. Colleoni
.594		Teatro Sociale	Giovanni Battista Vergnani	1842				Lombardia	Mantova	Bozzolo	Piazza Europa
.595		Teatro Sociale	Arnaldo Trebeschi	1853	1873	1954		Lombardia	Brescia	Brescia	Via Felice Cavallotti, 20
.596		Teatro Sociale	Salvatore Bruno	1876	1930			Liguria	Genova	Camogli	Piazza Giacomo Matteotti
.597		Teatro Sociale	Ernesto Basile	1908	2009			Sicilia	Agrigento	Canicatti	Via Capitano Ippolito, 5
.598		Teatro Sociale		1920	2006			EmiliaRomagna	Reggio Emilia	Canossa	Piazza Matilde di Canossa, 2
.599		Teatro Sociale	Gaetano Besia	1829	1985			Lombardia	Como	Canzo	Via Alessandro Volta, 2
.600		Teatro Sociale		1783				Lombardia	Cremona	Casalmaggiore	Via Cairoli

Site of arch. int.	UNESCO Site	ERHT Site	Integrated/Isolated	Construction	Seats	Level	Keyword	Notes	Code
Yes	No	No	Isolated	Wooden and brick		Local			.591
Yes	No	No	Isolated	Wooden and brick	165	Local			.592
Yes	No	No	Isolated	Mixed structure	1300	national			.593
Yes	No	No	Isolated	Wooden and brick		Local			.594
Yes	No	No	Isolated	Mixed structure		Regional		Rebuilt twice	.595
Yes	No	No	Isolated	Mixed structure		Local			.596
Yes	No	No	Isolated	Reinforced concrete	300	Regional			.597
Yes	No	No	Isolated	Mixed structure	99	Local			.598
Yes	No	No	Isolated	Mixed structure		Local			.599
Yes	No	No	Isolated	Wooden and brick	368	Regional			.600

List of Historical Theatres in Italy

Code	Code2	Name	Architect/Engineer	Year	R1	R2	R3	Region	Province	City	Address
.601		Teatro Sociale	Luigi Canonica	1843				Lombardia	Mantova	Castiglione delle Stiviere	Via Teatro, 13
.602		Teatro Sociale		1828				Veneto	Padova	Cittadella	Via Indipendenza
.603		Teatro Sociale	Canonica, Cusi	1831				Lombardia	Como	Como	Via Bellini, 3
.604		*Teatro Sociale*		*1732*	*1782*	*1838*	*1929*	*Lombardia*	*Cremona*	*Crema*	*Piazza Guglielmo Marconi, 17*
.605		Teatro Sociale		1867				Puglia	Brindisi	Fasano	Via Nazionale dei Trulli
.606		Teatro Sociale		1862				Lombardia	Milano	Gallarate	Via Teatro, 5
.607		Teatro Sociale	Giuseppe Erbesato	1811				Veneto	Verona	Isola della Scala	Via C. Battisti, 9B
.608		Teatro Sociale		1900				Lombardia	Varese	Luino	Corso 25 Aprile 1945, 13
.609		*Teatro Sociale*	*Luigi Canonica*	*1817*	*1845*	*1888*	*2014*	*Lombardia*	*Mantova*	*Mantova*	*Piazza Cavallotti, 14*
.610		Teatro Sociale		1890				Lombardia	Sondrio	Mobergno	Piazza Enea Mattei, 1

Site of arch. int.	UNESCO Site	ERHT Site	Integrated/Isolated	Construction	Seats	Level	Keyword	Notes	Code
Yes	No	No	Isolated	Wooden and brick		Local			.601
Yes	No	No	Isolated	Wooden and brick	470	Regional			.602
Yes	No	No	Isolated	Wooden and brick	999	Regional			.603
Yes	*No*	*No*	*Isolated*	*Mixed structure*		*Local*		*Enlarged by Piermarini in 1782, demolished due to a fire in 1937*	*.604*
Yes	No	No	Isolated	Wooden and brick	240	Local			.605
Yes	No	No	Isolated	Wooden and brick	500	Regional			.606
Yes	No	No	Isolated	Wooden and brick		Local			.607
No	No	No	Isolated	Wooden and brick	497	Regional			.608
Yes	*Yes*	*Yes*	*Isolated*	*Mixed structure*	852	*Regional*	*Urban node*		*.609*
Yes	No	No	Isolated	Wooden and brick	424	Regional			.610

List of Historical Theatres in Italy

Code	Code2	Name	Architect/Engineer	Year	R1	R2	R3	Region	Province	City	Address
.611		Teatro Sociale	Tosi, De Carlo	1920	1983			EmiliaRomagna	Rimini	Novafeltria	Via Giuseppe Mazzini, 69
.612		Teatro Sociale	Pietro Pivi	1926				EmiliaRomagna	Modena	Novi di Modena	Via Martiri della Libertà, 2
.613		Teatro Sociale		1902		1997		Piemonte	Verbano-Cusio-Ossola	Omegna	Via Carducci, 2
.614		Teatro Sociale		1870	1971			Lombardia	Brescia	Palazzolo sull'Oglio	Piazza Zamara, 9
.615		Teatro Sociale		1818				Lombardia	Brescia	Pontevico	Vicolo del Teatro
.616		Teatro Sociale	Sante Basseggio	1819				Veneto	Rovigo	Rovigo	Piazza Garibaldi, 14
.617		Teatro Sociale	Achille Sfondrini	1873	2002			Lombardia	Brescia	Salò	Via S. Bernardino, 48
.618		Teatro Sociale	Luigi Canonica	1824	2014			Lombardia	Sondrio	Sondrio	Piazza Garibaldi Giuseppe, 26
.619		Teatro Sociale		1840				Lombardia	Cremona	Soresina	Via Giuseppe Verdi, 23
.620		Teatro Sociale	Giovanbattista Chiappa	1846	1910			Lombardia	Pavia	Stradella	Via Faravelli, 2

Site of arch. int.	UNESCO Site	ERHT Site	Integrated/Isolated	Construction	Seats	Level	Keyword	Notes	Code
Yes	No	No	Isolated	Mixed structure	239	Local			.611
Yes	No	No	Isolated	Reinforced concrete	140	Local		Closed in 1980, damaged by the 2012 earthquake	.612
No	No	No	Isolated	Reinforced concrete	426	Regional			.613
Yes	No	No	Isolated	Mixed structure		Local			.614
Yes	No	No	Isolated	Wooden and brick		Local			.615
Yes	No	No	Isolated	Wooden and brick		Regional			.616
Yes	No	No	Isolated	Mixed structure		Local			.617
Yes	No	No	Isolated	Mixed structure	580	Regional			.618
Yes	No	No	Isolated	Wooden and brick		Local			.619
Yes	No	No	Isolated	Mixed structure	300	Regional			.620

List of Historical Theatres in Italy

Code	Code2	Name	Architect/Engineer	Year	R1	R2	R3	Region	Province	City	Address
.621		Teatro Sociale	Giuseppe Maria Ducati	1819				TrentinoAltoAdige	Trento	Trento	Via Paolo Mazzurana
.622		Teatro Sociale	Pietro Clerici	1861	1923	200		Piemonte	Alessandria	Valenza	Corso Garibaldi, 58
.623		Teatro Sociale		1931				Piemonte	Vercelli	Vercelli	Via Monte di Pietà, 15
.624		Teatro Sociale		1911				Veneto	Verona	Villa Bartolomea	Corso Fraccaroli, 54
.625		Teatro Sociale		1911				Lombardia	Mantova	Villastrada	Via XX Settembre
.626		Teatro Sociale	Giuseppe Segusini	1843	1925	2002		Veneto	Treviso	Vittorio Veneto	Via Martiri della Libertà, 36
.627		Teatro Sociale	Gioacchino Dell'Isola	1845				Lombardia	Pavia	Voghera	Via Emilia, 79

(continued)

(continued)

Code	Code2	Name	Architect/Engineer	Year	R1	R2	R3	Region	Province	City	Address
.628		Teatro Sociale di Finale Emilia	Giorgi, Finetti	1907	1990			EmiliaRomagna	Modena	Finale Emilia	Via Trento Trieste, 15
.629		Teatro Sociale Gaetano Bonoris		1774	1890	2003		Lombardia	Brescia	Montichiari	Piazza Teatro, 10
.630		Teatro Sociale Gian Marco Arrigoni		17th cent				FriuliVeneziaGiulia	Pordenone	San Vito al Tagliamento	Piazza del Popolo, 13

Site of arch. int.	UNESCO Site	ERHT Site	Integrated/Isolated	Construction	Seats	Level	Keyword	Notes	Code
Yes	No	No	Isolated	Wooden and brick	754	Regional			.621
Yes	No	No	Isolated	Mixed structure	98	Local			.622
No	No	No	Isolated	Reinforced concrete	800	Regional		Reconstruction of the previous theatre of 1815 destroyed by a fire in 1923	.623
Yes	No	No	Isolated	Mixed structure	650	Regional			.624
Yes	No	No	Isolated	Mixed structure		Local			.625
Yes	No	No	Isolated	Mixed structure	394	Regional			.626
Yes	No	No	Isolated	Wooden and brick		Local		Closed due to unavailability in 1985, to be restored	.627
Yes	No	No	Isolated	Reinforced concrete	500	Regional			.628
Yes	No	No	Isolated	Mixed structure	375	Local			.629
No	No	No	Isolated	Wooden and brick		Regional			.630

Code	Code2	Name	Architect/Engineer	Year	R1	R2	R3	Region	Province	City	Address
.631		Teatro Sociale Giorgio Busca	Giorgio Busca	1855				Piemonte	Cuneo	Alba	Piazza Vittorio Veneto, 3
.632		Teatro Sociale Soms		1900				Lombardia	Mantova	Castellucchio	Strada Provinciale 55, 32
.633		Teatro Sociale Stabile del Veneto Carlo Goldoni		1622	1979			Veneto	Venezia	Venezia	S. Marco, 4650/B
.634		Teatro Sociale Villani	Giuseppe Bollati	1894	1935	1950	2002	Piemonte	Biella	Biella	Piazza Martiri della Libertà, 2
.635		Teatro Solvay		1928				Toscana	Livorno	Rosignano Solvay	Via E. Solvat, 20
.636		Teatro Stabile Sloveno	Mihevc, Rusanov, Rozman	1902				FriuliVeneziaGiulia	Trieste	Trieste	Via Petronio, 4
.637		Teatro Storchi	Vincenzo Maestri	1889	1929	1986		EmiliaRomagna	Modena	Modena	Largo Garibaldi, 15

(continued)

(continued)

Code	Code2	Name	Architect/ Engineer	Year	R1	R2	R3	Region	Province	City	Address
.638		Teatro Subasio		1780	1799	1880	1995	Umbria	Perugia	Spello	Via Salnitraria, 15
.639		Teatro Sybaris		1845				Calabria	Cosenza	Cosenza	Contrada Calandrino, 14C
.640		Teatro Talia Marsica		1832	2000			Abruzzo	L'aquila	Tagliacozzo	Via G. Matteotti

Site of arch. int.	UNESCO Site	ERHT Site	Integrated/Isolated	Construction	Seats	Level	Keyword	Notes	Code
Yes	No	No	Isolated	Wooden and brick	500	Regional			.631
No	No	No	Isolated	Wooden and brick	190	Local			.632
Yes	Yes	No	Isolated	Mixed structure	800	Regional			.633
Yes	No	No	Isolated	Mixed structure	630	Regional			.634
Yes	No	No	Isolated	Reinforced concrete	600	Regional			.635
Yes	No	No	Isolated	Mixed structure	543	Regional			.636
Yes	No	No	Isolated	Mixed structure	952	Regional			.637
Yes	No	No	Isolated	Mixed structure	400	Local			.638
Yes	No	No	Isolated	Wooden and brick		Regional			.639
Yes	No	No	Isolated	Mixed structure	207	Local			.640

List of Historical Theatres in Italy

Code	Code2	Name	Architect/Engineer	Year	R1	R2	R3	Region	Province	City	Address
.641		Teatro Tartaro		1930	2003			Puglia	Lecce	Galatina	Corso Principe di Piemonte, 19
.642		Teatro Toniolo	Francesconi, Fabbris	1913	2008			Veneto	Venezia	Mestre	Piazzetta Cesare Battisti, 1
.643	*CODE 5.01 and 5.21*	*Teatro Tor di Nona*	*Fontana, Giorgi, Morelli*	*1671*	*1695*	*1795*		*Lazio*	*Roma*	*Roma*	*Lungotevere Tor di Nona*
.644		Teatro Torrielli		1890				Piemonte	Alessandria	Ovada	Via Cairoli
.645		Teatro Torti		1886	1994			Umbria	Perugia	Bevagna	Via del Gonfalone, 2
.646		Teatro Traetta		1838	2002			Puglia	Bari	Bitonto	Largo Teatro, 17
.647		Teatro Traiano	Antonio De Rossi	1844	1948	1978		Lazio	Roma	Civitavecchia	Corso Centocelle, 2
.648		Teatro Triano Viviani		1911	2002			Campania	Napoli	Napoli	Piazza Vincenzo Calenda
.649		Teatro Trianon	Vincenzo Bacigalupi	1913				Liguria	La Spezia	La Spezia	Via Manzoni
.650		Teatro Trifiletti	Letterio Savoja	1912				Sicilia	Messina	Milazzo	Via Cumbo Borgia

Site of arch. int.	UNESCO Site	ERHT Site	Integrated/Isolated	Construction	Seats	Level	Keyword	Notes	Code
Yes	No	No	Isolated	Mixed structure	266	Local		Recovery and change of intended use, cinemas and shops	.641
Yes	No	No	Isolated	Reinforced concrete	728	Regional			.642
Yes	Yes	No	Isolated	Wooden and brick		National	Architectural node	Rebuilt in 1733 and 1781, portico by Valadier, demolished in 1888	.643
Yes	No	No	Isolated	Wooden and brick		Local		Closed due to unavailability	.644
Yes	No	No	Isolated	Mixed structure	251	Local			.645
Yes	No	No	Isolated	Mixed structure	288	Local			.646
Yes	No	No	Isolated	Mixed structure	638	Local		Damaged by the bombings of 1943 and rebuilt	.647
Yes	Yes	No	Isolated	Reinforced concrete	530	Regional			.648
Yes	No	No	Isolated	Mixed structure		Regional			.649
Yes	No	Yes	Isolated	Mixed structure		Local			.650

List of Historical Theatres in Italy

Code	Code2	Name	Architect/Engineer	Year	R1	R2	R3	Region	Province	City	Address
651		Teatro Tullio Serafin	Giò Piasenti	1878	2007			Veneto	Venezia	Cavarzere	Via Roma, 8
652		Teatro Turreno	Alessandro Arienti	1891	1990	2013		Umbria	Perugia	Perugia	Piazza Danti, 13
653		Teatro Umberto		1880				Piemonte	Alessandria	Ricaldone	Via Roma, 8
654		Teatro Unione	Virginio Vespignani	1855				Lazio	Viterbo	Viterbo	Via Teatro Nuovo
655		Teatro Valle	Tommaso Morelli	1727	1821	2016		Lazio	Roma	Roma	Via del Teatro Valle, 21
656		Teatro Van Westerhout	Vittorio Chiaia	1896	1908	1972	2000	Puglia	Bari	Mola di Bari	Via Van Westerhout, 17
657		Teatro Vecchio		1832				EmiliaRomagna	Parma	Bardi	Via Pietro Cella, 7
658		Teatro Vecchio		1868				Lombardia	Mantova	Gonzaga	Vicolo Teatro Vecchio
659	*CODE 4.21*	*Teatro Vecchio*	*Viani, Motta, Cadioli*	*1628*				*Lombardia*	*Mantova*	*Mantova*	
660		Teatro Velluti		1819	1992			Marche	Macerata	Corridonia	Piazza del Popolo, 1

Site of arch. int.	UNESCO Site	ERHT Site	Integrated/Isolated	Construction	Seats	Level	Keyword	Notes	Code
Yes	No	No	Isolated	Mixed structure		Local			.651
Yes	No	No	Isolated	Mixed structure	1200	National		Closed since 2010, change of intended use	.652
Yes	No	No	Isolated	Wooden and brick	100	Local			.653
Yes	No	No	Isolated	Wooden and brick	764	Regional			.654
Yes	Yes	No	Isolated	Mixed structure		Regional			.655
Yes	No	No	Isolated	Mixed structure	102	Local			.656
Yes	No	No	Isolated	Wooden and brick		Local			.657
Yes	No	No	Isolated	Wooden and brick		Local			.658
Yes	*Yes*	*No*	*Integrated*	*Wooden and brick*		*Regional*	*Recast*		*.659*
Yes	No	No	Isolated	Mixed structure		Local			.660

List of Historical Theatres in Italy

Code	Code2	Name	Architect/Engineer	Year	R1	R2	R3	Region	Province	City	Address
.661		Teatro Ventidio Basso	Ireneo Aleandri	1846				Marche	Ascoli Piceno	Ascoli Piceno	Via del Trivio
.662		Teatro Verdi	Pier Luigi Montecchini	1868	2000			EmiliaRomagna	Parma	Busseto	Piazza Verdi, 24
.663		Teatro Verdi		1913	2012			Toscana	Pisa	Casciana Terme	Viale Regina Margherita, 13
.664		Teatro Verdi		1874				EmiliaRomagna	Forli-Cesena	Cesena	Via Luigi Sostegni, 13
.665		Teatro Verdi		1900				Veneto	Vicenza	Costabissara	Piazza Vittorio Veneto, 29
.666		*Teatro Verdi*	*Antonio Foschini*	*1913*	*2011*			*EmiliaRomagna*	*Ferrara*	*Ferrara*	*Via X Martiri, 141*
.667		Teatro Verdi		1853	2006			EmiliaRomagna	Piacenza	Fiorenzuola d'Arda	Via Liberazione
.668		Teatro Verdi	Telemaco Bonaiuti	1854	1950	1985		Toscana	Firenze	Firenze	Via Ghibellina, 99
.669		Teatro Verdi	Giacomo Fabbri	1882	1934	1982		EmiliaRomagna	Forli-Cesena	Forlimpopoli	Piazza Fratti, 7/8
.670		Teatro Verdi		1781	1846	1938	2002	FriuliVeneziaGiulia	Gorizia	Gorizia	Via Garibaldi, 2A

Site of arch. int.	UNESCO Site	ERHT Site	Integrated/Isolated	Construction	Seats	Level	Keyword	Notes	Code
Yes	No	No	Isolated	Wooden and brick	842	Regional			.661
Yes	No	No	Isolated	Mixed structure	307	Local			.662
Yes	No	No	Isolated	Reinforced concrete	324	Regional			.663
Yes	No	No	Isolated	Wooden and brick	600	Regional			.664
No	No	No	Isolated	Mixed structure		Local			.665
Yes	*Yes*	*No*	*Isolated*	*Mixed structure*	*1400*	*Regional*	*Organism*	*On the grounds of the Teatro dell'Accademia degli Intrepidi, then Teatro Obizzi, then Teatro Tosi-Borghi*	*.666*
Yes	No	No	Isolated	Mixed structure		Local			.667
Yes	Yes	No	Isolated	Mixed structure	1538	National			.668
Yes	No	No	Isolated	Mixed structure	204	Local			.669
Yes	No	No	Isolated	Mixed structure		Regional		Reconstruction of the previous theatre of 1740 destroyed by a fire in 1779, restructuring of 1846 by Andrea Scala	.670

List of Historical Theatres in Italy

Code	Code2	Name	Architect/ Engineer	Year	R1	R2	R3	Region	Province	City	Address
.671		Teatro Verdi		1900				FriuliVeneziaGiulia	Pordenone	Maniago	Via Umberto I, 53
.672		Teatro Verdi	Lodovico Fortini	1829	1981			Toscana	Pistoia	Montecatini Terme	Via Giuseppe Verdi, 45
.673		Teatro Verdi		1777	1901	2000		Toscana	Arezzo	Monte San Savino	Via Sansovino, 66
.674		Teatro Verdi	Antonio Cugini	1751	1847	1920		Veneto	Padova	Padova	Via Livello, 32
.675		Teatro Verdi	Scala, Giardi	1867				Toscana	Pisa	Pisa	Via Palestro, 40
.676		Teatro Verdi		1900				Piemonte	Alessandria	Pontestura	Piazza Castello, 19
.677		Teatro Verdi	Cesare Bazzani	1819	1886			Puglia	Foggia	San Severo	Corso Garibaldi
.678		Teatro Verdi		1902				Toscana	Firenze	Santa Croce sull'Arno	Via Verdi
.679		Teatro Verdi Buscoldo		1913				Lombardia	Mantova	Curtatone	Via G. Marconi, 13
.680		Teatro Vincenzo Bellini		1779	1889	1933	2004	Sicilia	Catania	Adrano	Piazza Duca degli Abruzzi, 23

Site of arch. int.	UNESCO Site	ERHT Site	Integrated/Isolated	Construction	Seats	Level	Keyword	Notes	Code
No	No	No	Isolated	Wooden and brick	420	Regional			.671
Yes	No	No	Isolated	Mixed structure	1000	National			.672
Yes	No	No	Isolated	Mixed structure	230	Local			.673
Yes	No	No	Isolated	Mixed structure	700	Regional			.674
Yes	No	No	Isolated	Wooden and brick	888	Regional			.675
No	No	No	Isolated	Mixed structure		Local			.676
Yes	No	No	Isolated	Mixed structure	430	Regional			.677
No	No	No	Isolated	Mixed structure	295	Local			.678
Yes	No	No	Isolated	Reinforced concrete	200	Local			.679
Yes	No	No	Isolated	Mixed structure		Local			.680

List of Historical Theatres in Italy

Code	Code2	Name	Architect/Engineer	Year	R1	R2	R3	Region	Province	City	Address
.681		Teatro Vincenzo Pagani	Antonio Murri	1875				Marche	Fermo	Monterubbiano	Via Trento Trieste, 1
.682		Teatro Vittoria		1923				EmiliaRomagna	Rimini	Pennabili	Piazza G. B. Mastini
.683		Teatro Vittorio Alfieri		1872	2017			Sicilia	Messina	Naso	Piazza Giuseppe Garibaldi, 11
.684		Teatro Vittorio Emanuele	Telemaco Buonaiuti	1862	1932			Toscana	Firenze	Firenze	Corso Italia, 16
.685		Teatro Vittorio Emanuele	Pietro Valente	1852	1980			Sicilia	Messina	Messina	Via Pozzo Leone, 5
.686		Teatro Yves Montand		1896	1928	2006		Toscana	Pistoia	Monsummano Terme	Piazza del Popolo, 97
.687		Teatro Zago		1891				Veneto	Rovigo	Loreo	piazza San Pietro
.688		Teatro Zanardelli		1890				Lombardia	Brescia	Gottolengo	Piazza XX Settembre, 8,
.689		Teatro Zancanaro		1910	1997			FriuliVeneziaGiulia	Pordenone	Sacile	Viale Zancanaro Pietro, 26
.690		Teatro Zandonai		1785	1828	1872		TrentinoAltoAdige	Trento	Rovereto	Corso Bettini, 82

Site of arch. int.	UNESCO Site	ERHT Site	Integrated/Isolated	Construction	Seats	Level	Keyword	Notes	Code
Yes	No	No	Isolated	Wooden and brick	260	Local			.681
Yes	No	No	Isolated	Reinforced concrete	149	Local			.682
Yes	No	No	Isolated	Mixed structure		Local			.683
Yes	No	No	Isolated	Mixed structure	1800	National			.684
Yes	No	No	Isolated	Mixed structure	2212	National		Rebuilt after the 1908 earthquake	.685
Yes	No	No	Isolated	Mixed structure	316	Local			.686
Yes	No	No	Isolated	Mixed structure		Local			.687
Yes	No	No	Isolated	Wooden and brick		Local			.688
Yes	No	No	Isolated	Mixed structure		Local			.689
Yes	No	No	Isolated	Mixed structure	780	Regional			.690

References

(1980) Il potere e lo spazio. La scena del principe. Electa, Firenze

(1989) Il teatro a Roma nel Settecento. Istituto dell'Enciclopedia italiana, Roma

Allegri L (1988) Teatro e spettacolo nel Medioevo. Editori Laterza, Roma

Baker HB (1889) The London stage. Allen & Co., Londra

Baur-Heinhold M (1971) Teatro Barocco. Electa, Milano

Breton G (1990) Teatri. tecniche nuove, Milano

Cheney S (1917) The Art Theatre. A.A. Knopf, New York

Cheney S (1918) The open-air theatre. Mitchell Kennerley, New York

Cheney S (1919) The theatre. Tudor Publishing Company, New York

Ciancarelli F (1987) Il progetto di una festa barocca. Alle origini del Teatro Farnese di Parma (1618–1629). Bulzoni Editore, Roma

Corsi M (1939) Il teatro all'aperto in Italia. Rizzoli, Roma

Cruciani F (1983) Teatro nel Rinascimento. Roma 1450–1550. Bulzoni editore, Roma

Cruciani F (1992) Lo spazio del teatro. Laterza, Bari

Howard D, Moretti L (2006) Architettura e musica nella Venezia del Rinascimento. Mondadori, Milano

Izenour CG (1977) Theater design. Mcgraw Hill, London

Izenour CG (1988) Theater technology. Mcgraw Hill, London

Izenour CG (1992) Roofed theatres of classical antiquity. Yale University Press, London

Konigson E (1990) Lo spazio del teatro nel medioevo. La Casa Usher, Lucca

Landriani P (1830) Storia e descrizione de' principali teatri antichi e moderni corredata di tavole. Giulio Ferrario Editore, Milano

Magagnato L (1954) Teatri italiani del Cinquecento. Neri Pozza Editore, Venezia

Mangini N (1974) I teatri di Firenze. Mursia editore, Milano

Milesi F (2000) Giacomo Torelli. L'invenzione scenica nell'Europa barocca. Fondazione Cassa di Risparmio di Fano, Fano

Montella C (2007) Mies van der Rohe e il teatro del Novecento: il teatro di Mannheim: dall'illusione all'astrazione. Electa, Napoli

Nagler AM (1973) Shakespeare's Stage. Yale University Press, New Haven

Narpozzi M (2006) *Teatri*. Motta Architettura, Milano

Pevsner N (1976) A history of buiding types. Thames and Hudson, Londra

Pieri M (1989) La nascita del teatro moderno. Bollati Boringhieri, Torino

Ragghianti CL (1952) Cinema arte figurativa. Einaudi, Torino

Rava A (1940) Teatro medievale. Coletti, Roma

Ricci C (1915) I Bibiena architetti teatrali. Alfieri & Lacroix, Milano

Ricci G (1971) Teatri d'Italia. Bramante Editrice, Milano

© Springer Nature Switzerland AG 2022
S. Clemente, *Reconstructing Theatre Architecture*, The Urban Book Series,
https://doi.org/10.1007/978-3-030-89968-4

364 References

Ronconi L (1992) Lo spettacolo e la meraviglia: il teatro Farnese di Parma e la festa barocca. Nuova ERI, Torino
Rosselli P, Romby GC, Fantozzi Micali O (1978) I teatri di Firenze. Bonechi, Firenze
Schnapper A (1982) La scenografia barocca. Editrice CLUEB, Bologna
Strappa G (1995) Unità dell'organismo architettonico. Edizioni Dedalo, Bari
Strappa G (2014) L'architettura come processo. Il mondo plastico murario in divenire. Franco Angeli, Milano

Ancient treatises

Sabatini N (1638) Pratica di fabricar scene, e machine ne' teatri. P. de' Paoli e G.B. Giovannelli Stampatori Camerali, Ravenna
Troili G (1683) Paradossi per praticare la prospettiva senza saperla. G. Longhi, Bologna
Maffei S (1753) De' teatri antichi, e moderni. Agostino Carattoni, Verona
Riccati F (1790) Della costruzione de' teatri secondo il costume d'Italia. Remondini di Venezia, Bassano
Milizia F (1794) Trattato completo, formale e materiale del Teatro. P. e G.B. Pasquali, Venezia
Giorgi F (1795) Descrizione istorica del teatro di Tor di Nona. Cannetti, Roma
Sabbatini N (1955) Pratica di fabricar scene e machine ne' teatri. Bestetti, Roma
Carini Motta F (1972) Trattato sopra la struttura de' teatri e scene. Il profilo, Milano
Barbaro D (1980) La pratica della perspettiva. Forni, Bologna
Scamozzi V (1997) L'Idea dell'Architettura Universale. Centro Internazionale di Studi Andrea Palladio, Vicenza
Bastianello E (2015) Pellegrino Prisciani. Spectacula. Guaraldi Engramma, Rimini

List of the main archives and libraries

Accademia di Danimarca
American Academy
Archivio del Teatro dell'Opera di Roma
Archivio di Stato di Mantova
Archivio di Stato di Milano
Archivio di Stato di Roma
Archivio di Stato di Venezia
Archivio Storico Comunale di Ferrara
Archivio Storico Comunale di Mantova
Archivio storico del Teatro alla Scala
Biblioteca Comunale Ariostea
Biblioteca dell'Istituto Archeologico Germanico di Roma
Bibliotheca Hertziana
Biblioteca teatrale SIAE del Burcardo
Biblioteca Teresiana
BSR library
Centro Archivi di Architettura, MAXXI
Theatre Institute's library

References

"There is nothing in all of Europe that I don't say comes close to this theater, but gives a clue. The eyes are dazzled, the soul kidnapped".
Stendhal.